Biogenesis of Plant Cell Wall Polysaccharides

ACADEMIC PRESS RAPID MANUSCRIPT REPRODUCTION

Proceedings of a Symposium on
Biogenesis of Plant Cell Wall Polysaccharides
Held at the 164th National Meeting of the
American Chemical Society, New York, New York
August 28-29, 1972

Biogenesis of Plant Cell Wall Polysaccharides

Edited by

Frank Loewus

Department of Biology
State University of New York
Buffalo, New York

Academic Press

NEW YORK AND LONDON

1973

A Subsidiary of Harcourt Brace Jovanovich, Publishers

ACADEMIC PRESS, INC.
111 Fifth Avenue, New York, New York 10003

United Kingdom Edition published by
ACADEMIC PRESS, INC. (LONDON) LTD.
24/28 Oval Road, London NW1

Library of Congress Cataloging in Publication Data
Main entry under title.

Biogenesis of plant cell wall polysaccharides.

"Proceedings of a symposium on biogenesis of plant
cell wall polysaccharides held at the 164th national
meeting of the American Chemical Society, New York,
New York, August 28-29, 1972."
Symposium sponsored by the Division of Cellulose,
Wood, and Fiber Chemistry of the American Chemical
Society.
1. Polysaccharide synthesis—Congresses. 2. Plant
cell walls—Congresses. I. Loewus, Frank Abel, DATE
ed. II. American Chemical Society.
III. American Chemical Society. Division of Cellulose,
Wood, and Fiber Chemistry.
QK898.P77B56 1973 581.8'75 72-9993
ISBN 0−12−455350−8

PRINTED IN THE UNITED STATES OF AMERICA

CONTENTS

CONTENTS

CONTRIBUTORS

Albersheim, Peter, Department of Chemistry, University of Colorado, Boulder, Colorado 80302

Aspinall, Gerald O., Department of Chemistry, York University, Downsview, Ontario, Canada

Bandurski, Robert S., Department of Botany and Plant Pathology, Michigan State University, East Lansing, Michigan 48823

Bauer, W. Dietz, MSU/AEC Plant Research Laboratory, Michigan State University, East Lansing, Michigan 48823

Brown, R. Malcolm, Jr., Department of Botany, University of North Carolina, Chapel Hill, North Carolina 27514

Cetorelli, J. J., Department of Biochemistry, J. Hillis Miller Health Center, University of Florida, Gainesville, Florida 32601

Chen, Min-Shen, Diabetes Center, Department of Medicine, New York Medical College, 1249 Fifth Avenue, New York, New York 10029

Chrispeels, Maarten J., Department of Biology, John Muir College, University of California at San Diego, La Jolla, California 92037

Colvin, J. Ross, Division of Biological Sciences, National Research Council of Canada, Ottawa, Canada

Davies, Michael D., Instituto de Biologia, Escola Superior de Agricultura, Universidade Federal de Vicosa, Vicosa, Minas Gerais, Brazil

Dickinson, David B., Department of Horticulture, University of Illinois, Urbana, Illinois 61801

Elbein, A. D., Department of Biochemistry, The University of Texas Medical School, San Antonio, Texas 78229

Fan, Der Fong, Montefiore Hospital, Pittsburgh, Pennsylvania 15213

Feingold, David Sidney, Department of Microbiology, School of Medicine, University of Pittsburgh, Pittsburgh, Pennsylvania 15213

Forsee, W. T., Department of Biochemistry, The University of Texas Medical School, San Antonio, Texas 78229

Franke, W. Werner, Division of Cell Biology, University of Freiburg, Freiburg i. Br., West Germany

Herth, Werner, Division of Cell Biology, University of Freiburg, Freiburg i. Br., West Germany

Hopper, James E., Department of Biochemistry, University of Washington, Seattle, Washington 98105

Keestra, Kenneth, Department of Biology, Massachusetts Institute of Technology, Cambridge, Massachusetts 02139

Kindel, Paul K., Department of Biochemistry, Michigan State University, East Lansing, Michigan 48823

Kjosbakken, J., Division of Biological Sciences, National Research Council of Canada, Ottawa, Canada

Kroh, M., Department of Botany, University of Nijmegen, Nijmegen, The Netherlands

Labarca, C., Facultad de Ciencias, Universidad de Chile, Casilla 653, Santiago, Chile

Lamport, Derek T. A., MSU/AEC Plant Research Laboratory, Michigan State University, East Lansing, Michigan 48823

Leppard, Gary G., Water Science Subdivision, Inland Waters Directorate, Department of the Environment, Ottawa, Canada

Loewus, F., Department of Biology, State University of New York at Buffalo, Buffalo, New York 14214

Loewus, Mary W., Department of Biology, State University of New York at Buffalo, Buffalo, New York 14214

Piskornik, Z., Plant Physiology Department, Agricultural University, Cracow, Poland

CONTRIBUTORS

Ramus, J., Department of Biology, Yale University, New Haven, Connecticut 06520

Roberts, R. M., Department of Biochemistry, J. Hillis Miller Health Center, University of Florida, Gainesville, Florida 32601

Romanovicz, Dwight, Department of Botany, University of North Carolina, Chapel Hill, North Carolina 27514

Sadava, D., Department of Biology, John Muir College, University of California at San Diego, La Jolla, California 92037

Talmadge, Kenneth W., Department of Biological Sciences, Princeton University, Princeton, New Jersey 08540

PREFACE

This volume comprises papers presented at a symposium on the *Biogenesis of Plant Cell Wall Polysaccharides* which was sponsored by the Division of Cellulose, Wood, and Fiber Chemistry, of The American Chemistry Society, held in New York, New York, August 28-29, 1972.

Thanks to the cooperation and interest of all contributors, the entire program is contained in these pages. Its authors represent a broad range of interests that stretch from structural to functional aspects of cell wall polysaccharide biosynthesis in algae as well as higher plants. Within these chapters the reader will find a detailed account of current progress and understanding regarding the biosynthesis of cell wall components and the assembly of these components in the wall.

Two excellent reviews on the role of sugar nucleotides in polysaccharide biosynthesis (H. Nikaido and W. Z. Hassid) and the biogenesis of plant cell walls (G. D. McGinnis and F. Shafizadeh) have appeared in *Advances in Carbohydrate Chemistry and Biochemistry,* Volume 26, 1971. The reader may wish to refer to these comprehensive articles for information not presented in the papers of this symposium.

Grateful acknowledgement is made to Dr. Stanley Rowland, Program Chairman of the Division of Cellulose, Wood, and Fiber Chemistry, for proposing this symposium and assisting in its arrangement. Costs of attendance were borne by individual contributors and this expression of cooperation is deeply appreciated. Manuscripts were retyped with facilities provided by the Faculty of Natural Sciences and Mathematics, State University of New York at Buffalo. The typists were Miss Judith Lubkowski and Miss Lynne Heinz and deep appreciation is expressed for their skillful assistance.

F. Loewus

Biogenesis of Plant Cell Wall Polysaccharides

THE *MYO*-INOSITOL OXIDATION PATHWAY TO CELL WALL POLYSACCHARIDES*

F. Loewus, Min-Shen Chen and Mary W. Loewus
Department of Biology
State University of New York at Buffalo   14214

Abstract

Conversion of hexose to uronic acid and pentose units of cell wall polysaccharides in higher plants may occur by one or the other (or both) of the following pathways:

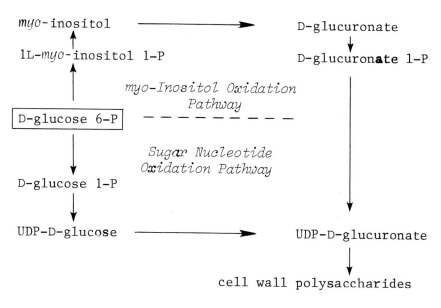

*This investigation was supported by NIH research grant number GM-12422 from the National Institute of General Medical Sciences.

1

Studies have been undertaken which seek to determine if the *myo*-inositol oxidation pathway performs a functional role during pectin biosynthesis. *myo*-Inositol is rapidly converted to pectic polysaccharides of pollen tube walls during germination of lily pollen. Experimental conditions for studying this conversion have been determined and applied to inhibitor studies involving the use of 2-*O*,*C*-methylene-*myo*-inositol (MMO). MMO blocks conversion of *myo*-inositol to uronic acid and pentose units of pectin but does not affect incorporation of hexose. When germinating pollen which has been pretreated with MMO is given to labeled D-glucose, net incorporation of label into polysaccharides of cytoplasm and tube wall is unchanged compared to untreated pollen but further examination of the distribution of label among monomeric residues reveals a significant decrease in labeled uronic acid and pentose units. Results indicate that under experimental conditions used, at least 70% of the glucose is converted to uronic acid and pentose products proceeds via the *myo*-inositol pathway.

The first enzyme of the *myo*-inositol oxidation pathway, D-glucose 6-P cycloaldolase, has been partially purified from cultured *Acer pseudoplatanus* cells. It contains bound NAD and exhibits properties resembling Type II aldolase.

Symbols

myoI	=	*myo*-inositol
myoIOP	=	*myo*-inositol oxidation pathway
DHAP	=	dihydroxyacetone phosphate
MMO	=	2-*O*,*C*-methylene-*myo*-inositol

Introduction

A similarity between products of oxidative *myo*-inositol metabolism and those of D-glucuronic acid metabolism in higher plants was reported in 1962 (1). This observation provided the experimental

technique needed to demonstrate that the biosynthesis
of myoI involved cyclization of D-glucose (2), thus
confirming a hypothesis made many years earlier by
Maquenne (3). On the basis of these studies, it was
proposed that a second pathway from D-glucose to
UDP-D-glucuronic acid, one that bypassed UDP-D-glucose dehydrogenase, may be operative in higher plants
(4). The two are outlined in Fig. 1. An important

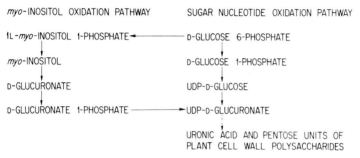

Fig. 1 Conversion of D-glucose 6-phosphate to
UDP-D-glucuronic acid and products of glucuronic
acid metabolism over the *myo*-inositol oxidation pathway and the sugar nucleotide pathway.

---

contribution to the idea of this alternative pathway
which for convenience is called the *myo*-inositol oxidation pathway, was the discovery that D-glucose
6-P is the substrate cyclized and 1L-*myo*-inositol
1-P is the product of this cyclization (5)*. Thus
D-glucose 6-P is common to both the myoIOP and the
sugar nucleotide oxidation pathway.

D-glucose 6-P cycloaldolase, the first enzyme of
the myoIOP, is found in plants, animals, microorgan-
isms (7, 8). During first steps of enzyme purifica-

---

*A description of current cyclitol nomenclature
is given in reference 6.

technique needed to demonstrate that the biosynthesis
tion, this cycloaldolase is accompanied by a phosphatase which selectively hydrolyzes equatorial
monophosphate esters of myoI (9). Fig. 1 contains
this phosphatase step and includes free myoI as an
intermediate. The possibility also exists that myoI
monophosphate or a bound form of myoI as yet unknown
functions at this point in the pathway as substrate
for D-glucuronic acid formation.

For the conversion of free myoI to UDP-D-glucuronic acid, 3 enzymes are required, an oxygenase, a
kinase and a pyrophosphorylase. The oxygenase has
been obtained as a cell-free preparation (10) but
never in purified form from plants. Information regarding its presence in plant tissues and its mode
of action has been obtained largely from tracer
studies which show that the site of oxidation is the
C1-C6 bond of myoI and the product is D-glucuronic
acid (11). Feingold, Neufeld and Hassid isolated
the kinase and pyrophosphorylase from mung bean
seedlings years ago (12). Recently, Roberts obtained
a purified pyrophosphorylase from barley seedlings
(13) and described its properties. A report of his
most recent work is included elsewhere in this volume.

Tracer studies have had a major role in estab-
lishing the stereochemistry of D-glucose cyclization
and myoI oxidation. Information drawn from results
of these studies is summarized in Fig. 2. Here,
the fate of specifically labeled D-glucose is traced
from D-glucose 6-P to 1L-*myo*-inositol 1-P to myoI.
Specifically labeled myoI is traced through open
chain and pyranose forms of D-glucuronic acid to
D-galacturonosyl residues of pectin. Observations
are as follows:

1. D-Glucose labeled C1, C2 or C6 with ^{14}C is con-
converted to myoI labeled in C6, C5, or C1, respectively (14).

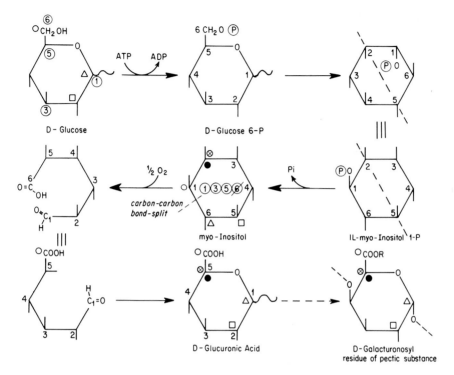

Fig. 2 Conversion of specifically labeled D-glucose to *myo*-inositol and conversion of specifically labeled *myo*-inositol to D-glucuronic acid and D-galacturonosyl residues of pectin. Specific carbon atoms labeled ^{14}C are marked $\triangle$, $\square$, $\bigcirc$ and $\bullet$ Carbon-bound tritium is marked with a $\otimes$ or with an open circle bearing the carbon number of the glucose position from which it was derived.

2. myoI labeled in C6, C5, C2, or C1 with ^{14}C is converted to uronic acid labeled in C1, C2, C5, or C6, respectively (1, 2, 14).

3. myoI labeled at C2 with tritium is converted to uronic acid labeled at C5. Free D-xylose, another labeled product formed in the course of this experiment, has all of its tritium at C5 stereospecifically located in the R position (15).

4.   D-Glucose labeled at Cl, C3, C5 or C6 with
tritium is converted to myoI with retention of
tritium.   In the case of D-glucose-6-^{3}H, only one-
half of the tritium appears in the product, corresponding to one of the two hydrogens attached to C6
of glucose (16, 17).

5.   Phosphate attached to D-glucose 6-P does not
migrate nor exchange during the conversion of 1L-
myo-inositol 1-P (17).

These findings, which represent the results of
many investigations in other laboratories as well
as ours, inform us that conversion of D-glucose to
D-glucuronic acid and its products via the myoIOP
will have the same relationship as regards carbon
chain sequence as that predicted via the sugar
nucleotide pathway.   In other words, specifically
labeled D-glucose-^{14}C or its 6-P will give no direct
information regarding the pathway of conversion involved in uronic acid biosynthesis.   This fact which
was fully appreciated in the earliest report (2),
prompted us to concentrate research efforts in 3
separate areas, each converging on our prinicpal
objective, the functional nature of the myoIOP.
These areas of research are:

a.   myoI metabolism, expecially those aspects lead-
      ing to cell wall polysaccharide biosynthesis,
b.   isolation and purification of enzymes implicated
      in the myoIOP, and
c.   studies which will modify the contribution of
      one or the other pathway to UDP-D-glucuronic
      acid formation.

myo-Inositol Metabolism

Conversion of myoI to uronic acid and pentose
residues of cell wall polysaccharides has been examined in several plant tissues (1, 2, 4, 11, 18, 19,

20). A summary of these findings is presented in Table 1. In some studies, myoI was supplied as a

Table 1

Components of Cell Wall Polysaccharides Derived from myo-Inositol

Component	Plant Tissue
D-Galacturonic acid	Strawberry fruit, parsley leaf, corn seedling, barley seedling, cultured Acer and Fraxinus cells, Avena coleoptiles, lily flowers, pollen tubes of lily and pear.
D-Glucuronic acid	Strawberry fruit, corn and barley seedlings.
4-O-Methyl-D-glucuronic acid	Corn and barley seedlings.
D-Xylose	Strawberry fruit, parsley leaf, corn and barley seedlings, cultured Acer cells, lily flowers and pollen tubes.
L-arabinose	Strawberry fruit, corn and barley seedlings, parsley leaf, cultured Acer cells, lily flowers, pollen tubes of lily and pear.

constituent of the medium, in others it was fed to detached plants or plant parts through the vascular tissue, and in still others it was injected directly into the plant part. Conversion of myoI to uronic acid and pentose residues of cell wall polysaccharides was obtained in each case. When myoI solution was applied directly to root hairs of germinating barley, as much as 80% of the applied label was found in cell wall polysaccharides (19). In recent studies of pollen tube wall formation, we have routinely obtained up to 70% conversion of the myoI supplied to *Lilium longiflorum* pollen (20). In the latter, there is evidence for an active system that transports myoI into growing pollen tubes.

7

D-Glucose 6-P Cycloaldolase, the First
Enzyme of the *myo*-Inositol Oxidation Pathway

D-Glucose 6-P is an intermediate common to
both the myoIOP and the sugar nucleotide oxidation
pathway. Beyond D-glucose 6-P, cycloaldolase is the
first enzyme of the myoIOP and as such may have important regulatory functions that influence both
hexose and myoI metabolism. Cycloaldolase has been
studied in several laboratories (5, 7-9, 16, 17, 21).
In ours, the enzyme has been isolated from a number
of plant species and, on the basis of this survey,
suspension cell culture of *Acer pseudoplatanus* was
chosen as a source of enzyme for further purification
and for studies of its enzymatic properties (8).

Unlike the enzyme from rat testis which becomes
less stable as purification proceeds, *Acer* enzyme is
quite stable and has been stored in its most purified form up to 3 months at 3 to 5 C with little
or no loss of activity. About 500 to 700 fold
purification is achieved in 3 steps: ammonium sulfate precipitation, DEAE-cellulose chromatography,
and gel filtration on Sephadex G-200. At this
stage the enzyme is still quite impure as seen by
disc gel electrophoresis (Fig. 3) but it is free of
phosphatase and hexose isomerase activities. The
activity is localized in zone 3 after gel electrophoresis, a densely staining double band of protein.
Gel permeation chromatography indicates a molecular
weight of 150,000.

Conversion of D-glucose 6-P to 1L-*myo*-inositol
1-P, the reaction catalyzed by cycloaldolase, and
dephosphorylation step that follows are given in
Fig. 4. Included are 2 proposed intermediates,
5-keto-D-glucose 6-P [also called D-*xylo*-5-hexulose
6-P] and *myo*-inosose-2 1-P [also called 2D-(2, 4, 6/
3, 5)-pentahydroxycyclohexanone 2-P]. The first step
in this reaction is considered to be a NAD$^+$-dependent
oxidation of C5 of D-glucose 6-P. In earlier studies

8

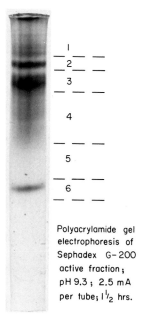

Polyacrylamide gel
electrophoresis of
Sephadex G-200
active fraction;
pH 9.3; 2.5 mA
per tube; $1\frac{1}{2}$ hrs.

Fig. 3 Disc gel electrophoresis of A. *pseudo-platanus* cycloaldolase after Sephadex G-200 chromatography of the enzyme.

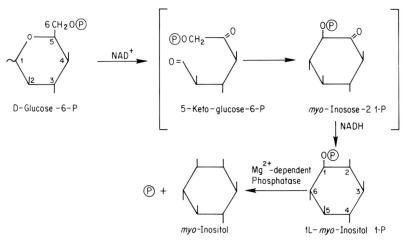

Fig. 4 Conversion of D-glucose 6-phosphate to *myo*-inositol by cycloaldolase and phosphatase. Postulated intermediates of the cycloaldolase reaction appear within the brackets.

9

we observed, as did others, an increase in the rate of reaction upon addition of NAD^+ and we assumed that this addition was essential. More recently, with shorter periods of dialysis and with reduced glutha-tione present at each step in purification up to Sephadex, we have obtained preparations which exhi-bit full activity in the absence of added NAD^+. Evidence of bound NAD^+ in these preparations has been obtained (M. W. Loewus and F. Loewus, *Plant Science Lett.*, in press). Prolonged dialysis of the enzyme with EDTA removes about 25% of the activity detect-able in the absence of added NAD^+. Addition of NAD^+ to dialyzed enzyme restores full activity. Oxidation of sulfhydryl groups on the enzyme with potassium tetrathionate followed by reversal of the reaction with dithiothreitol also lowers NAD^+ inde-pendent enzyme activity which can be fully restored by the addition of NAD^+.

At the time that cyclization of D-glucose to myoI was first described (2), it was suggested that a phosphorylated 5-keto-D-glucose might be an inter-mediate. More recently, indirect evidence for a reaction sequence involving this intermediate has been obtained from tracer studies in which deuterated or triatiated D-glucose 6-P was converted to 1L-*myo*-inositol 1-P or to free myoI (16, 17). Of particular interest are studies with D-glucose-5-^{2}H 6-P and D-glucose-5-^{3}H 6-P which show that the label is retained by the product, presumably at C2 although this point does not appear to have been examined experimentally. These studies indicate that the hydride ion generated by oxidation of C5 of D-glucose 6-P is transferred to NAD, that the reduced NAD remains bound to the enzyme during cyclization of the substrate, and that the same hydride ion is transfer-red back to the cyclized intermediate in a stereospeci-fic reduction. Sherman et al. (17) have shown that all intermediate products, not just NAD, are bound to the enzyme throughout the course of the reduction. In these respects, cycloaldolase resembles

that group of enzymes represented by UDP-D-galactose
4-epimerase, UDP-D-glucuronic acid lyase, dTDP-D-glu-
cose oxidoreductase, and UDP-D-apiose synthetase;
that is, an enzyme in which bound NAD functions
alternately as acceptor and then donor of a hydride ion during catalysis at the active site.

Cyclization, illustrated by conversion of 5-keto-
D-glucose 6-P to *myo*-inosose-2 1-P in Fig. 4, resembles the reaction catalyzed by aldolase. Preliminary studies (22) led us to suggest that cycloaldolase behaved like a Type I aldolase but further
work has revealed substantial evidence favoring
Type II aldolase-like activity. Evidence to support
the latter includes the following observations:

1.   EDTA inhibits the enzyme. The inhibition is
partially relieved by Mg^{2+} but not by Zn^{2+}, Co^{2+}, or
Mn^{2+}, presumptive evidence that the enzyme contains
a metal ligand.

2.   Potassium ions stimulate the enzyme.

3.   The enzyme has narrow pH optimum from 7.5 to 8.5.

4.   There is reason to believe that a lysine-bound
Schiff base intermediate, characteristic of Type I
aldolases is not formed during the reaction.

This last bit of information is based on studies
with DHAP which may be considered as an analog of
carbons 4, 5 and 6 of 5-keto-D-glucose 6-P. Kinetic
studies indicate that DHAP is a partially competitive
inhibitor of the enzyme (Fig. 5). In other words,
when the inhibitor has reached saturation, substrate
still interacts with the enzyme, undergoing oxidation,
cyclization and reduction to product. One explanation is that DHAP reacts with the same site as the
keto intermediate while D-glucose 6-P is bound at a
separate site and DHAP becomes competitive only after
the normal substrate is oxidized.

11

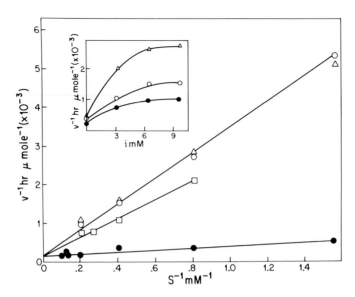

Fig. 5 Kinetics of inhibition of cycloaldolase by dihydroxyacetone phosphate. Symbols refer to: ●, control; □, 3.19 mM; ○, 6.38 mM; and △, 9.57 mM dihydroxyacetone phosphate. Inset: △, 1.24 mM; ○, 2.44 mM; and ●, 4.48 mM D-glucose 6-phosphate.

Reduction of the enzyme in the presence of DHAP with NaBH4 does not render the enzyme irreversibly inactive as was suggested in preliminary experiments (23). That observation appears to have stemmed from experiments lacking adequate controls, a condition now corrected. As seen in Table 2, permenent inactivation due to reduction of a Schiff base between DHAP and the enzyme was not observed. In fact, DHAP protects the enzyme against inactivation caused by NaBH4 reduction alone, an inactivation that may be caused by reduction of bound NAD$^+$, resulting in a reduced enzyme unable to participate in the reaction.

Table 2

Effect of NaBH$_4$ on cycloaldolase in the presence of DHAP

Additions	NaBH$_4$ added, μm		
	0	14	70
	enzyme activity, %		
Enzyme only, 19 μg	100	56	33
Enzyme + DHAP	86	100	51
Enzyme + DHAP + D-glucose 6-P	-	100	45

Absence of increased protection against NaBH$_4$ inactivation by a combination of DHAP and D-glucose 6-P would suggest that these 2 protective groups occupy separate sites, a point brought out earlier during discussion of DHAP inhibition.

The data obtained thus far (24) is consistent with the idea of an aldolase function resembling Type II aldolases which contain a metal ligand. When a purer enzyme becomes available, it will be most interesting to examine the regulatory function of cycloaldolase, not only in regard to its role in the myoIOP but also in regard to its interaction with other metabolic systems that intersect the pathway or compete for the substrate D-glucose 6-P.

Regarding the Functional Nature of the *myo*-Inositol Oxidation Pathway

There is mounting evidence to suggest that the myoIOP has functional significance in cell wall formation and that during certain stages of plant development it may be the major pathway for conversion of hexose to uronic acids and pentoses in plant polysaccharides. Some of this evidence is deduced from unsuccessful attempts to demonstrate

13

appreciable UDP-D-glucose dehydrogenase activity
in tissues undergoing rapid cell wall biosynthesis.
Some is derived from studies of competitive inhibition of UDP-D-glucose dehydrogenase by UDP-D-
glucuronic acid and UDP-D-galacturonic acid (25).
The functional role of the myoIOP can also be
examined by comparing relative contributions of
each of the alternate pathways from glucose to glucuronic acid, making use of the fact that myoI is
converted directly to glucuronic acid and products
of glucuronic acid metabolism while glucose is converted to products of hexose metabolism as well as
those of glucuronic acid. Use of this latter approach in our laboratory has uncovered experimental
evidence for the functional role of the myoIOP.

To meet the needs of this approach, a biological
system is needed in which both glucose and myoI are
readily converted to pectic substance in plant cell
walls. One such system, detached root tips of
germinating corn seedlings, has been used by R. M.
Roberts and F. Loewus (manuscript in preparation)
to explore the effect of raising the internal
concentration of myoI on the flow of carbon from
D-glucose-6-^{14}C to products of uronic acid metabolism
via the myoIOP. Results clearly indicate that it is
possible to reduce the contribution of ^{14}C from D-
glucose-6-^{14}C to cell wall galacturonic acid residues
by increasing the myoI level in the tissues of the
root tips.

Another system, *Lilium longiflorum* pollen, has
practical advantages which make it quite useful for
the kind of study described here. This pollen is
readily obtained, easily stored, and quite simple
to handle experimentally. The pollen will germinate
and produce pollen tubes in artificial media devoid
of a metabolizable carbon source yet it readily
utilizes specific carbon sources such as glucose and
myoI when these carbohydrates are added to the media.
The tubes are composed of primary cell wall exclusively and are rich in pectic substances.

14

Our studies with the *Lilium* pollen system were initiated by M. Kroh (26). Elsewhere in this volume Dr. Kroh presents an account of recent studies on cell wall substance in transmitting tissue of solid styles. Quite briefly, the assay with *Lilium* pollen consists of 1 ml of medium at pH 5.2 to which is added 5 mg of washed pollen grains and such labeled carbohydrate or chemical as one may wish to study. The medium consists of 0.3 M pentaerythritol, 1.3 mM calcium nitrate, 0.16 mM boric acid and 1 mM potassium nitrate. The pollen system is incubated with shaking in a 10 ml Erlynmeyer flask at 27 C for periods up to 10 hr. With longer periods pollen tubes grow to such lengths that they become entangled, creating problems in sample transfers and rinsing. Bacterial contamination must also be considered during long incubations.

Fig. 6 presents results on germination, tube length and incorporation of *myo*-inositol-2-^{3}H into pollen tube wall pectin over an 8 hr period in the absence of added compounds other than the standard media and 5 μg of labeled myoI and again with 10% of the medium replaced by undiluted stigmatic exudate from *L. longiflorum* pistils. Germination reached nearly 100% in both systems but took twice as long when exudate was deleted. Tube elongation was also accelerated by the presence of exudate. Incorporation of tritium into uronic acid and pentose units of pectic substance was also stimulated by the presence of exudate but the final level of incorporation, about 55%, was the same for both systems. These experiments provide a brief glimpse of the assay and its limitations. In the experiments to be described below, no exudate was present in the media (20).

Over a wide range of myoI concentration, uptake by pollen tubes was a linear function. Data in Fig. 7 was gathered after 4 hr of germination and growth in standard media containing the indicated

15

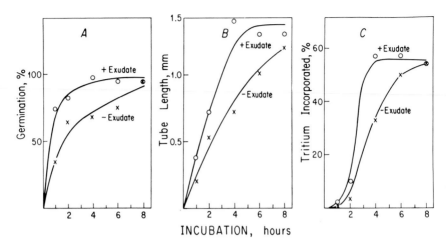

Fig. 6 Analysis of (A) germination, (B) pollen tube elongation, and (C) incorporation of tritium from *myo*-inositol-2-^{3}H into pollen tube wall polysaccharides in pollen of *L. longiflorum* cv. Croft incubated in pentaerythritol medium. See text for further details. Additional experimental details regarding results presented here and in Fig. 7–12 will be published separately.

myoI concentration. Changing myoI concentration did not affect germination but it did produce an increase in tube length (less than 10%) at 60 μg per ml or greater.

Our normal procedure for estimating incorporation of label into tube wall involves grinding pollen tubes in 70% ethanol followed by repeated washing of insoluble residues with more 70% ethanol until no more soluble radioactivity appears in the rinses. A more exacting procedure, one that leads to loss of all cold-water-soluble pectic substance, involves the grinding and extraction of labeled pollen tubes in water. Three such experiments are summarized in Fig. 8. At 50 μg of myoI per ml, 50% of the myoI label appeared in the tube wall fraction within 3 hr.

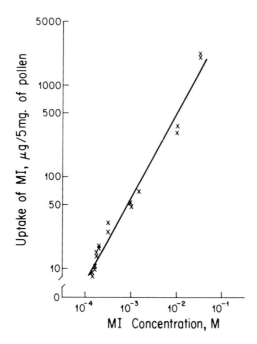

Fig. 7 Uptake of *myo*-inositol by germinating
L. longiflorum cv. Ace as a function of *myo*-inositol
concentration in pentaerythritol medium.

It should be emphasized that this is cold-water-
insoluble polysaccharide and as such is a minimal
estimate of incorporation. The decreasing rate
of incorporation of myoI into cytoplasm reflects
depletion of myoI in the medium. At higher levels
of myoI, 200 or 500 μg per ml, there was little
change in the overall rate of uptake, as seen by
comparing slopes of total uptake, but more myoI
label accumulated in the cytoplasm. Other studies
of the distribution of label, not presented here,
revealed that label continued to accumulate in the
membranous components of the cytoplasm and in the
70% ethanol soluble fraction of the cytoplasm
(after removing the membranous components) but accumulation in the 70% ethanol insoluble fraction
reached a plateau within 3 hr after pollen was placed

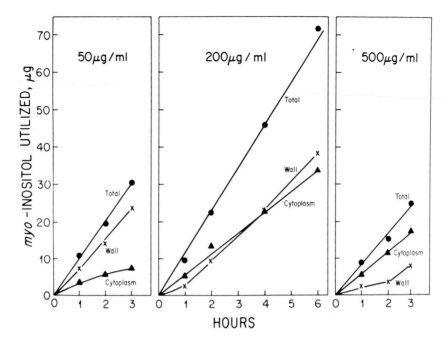

Fig. 8 Uptake of *myo*-inositol by *L. Longiflorum* pollen and incorporation of products of *myo*-inositol metabolism into pollen tube cytoplasm and wall polysaccharides. The pollen was pre-incubated for 3 hr in pentaerythritol medium prior to addition of *myo*-inositol.

in the labeled media. The latter probably corresponds to newly formed polysaccharide that has not been deposited at the wall, a small pool of carbohydrate that is constantly being turned over as growth proceeds. Pulse-chase studies should clarify this process. During the first 6 hr tube growth in artificial media, most of the labeled myoI is converted to D-galacturonic acid and L-arabinose residues of pectin. At longer intervals, label appears in xylose and hexose residues, the latter as a consequence of metabolic recycling of label derived from breakdown products of the primary conversion (26).

With conditions for tracing the conversion of
myoI to pectic substance of pollen tube wall well
in hand, attention was directed to the application
of the pollen system as a tool for testing the relative contributions of the 2 pathways to products
of uronic acid metabolism. The inhibitor chosen
for this study was the 2-epoxide of myoI, 2-0,C-
methylene-*myo*-inositol, a compound first used by
Schopfer and Posternak to study its effects on the
growth and morphology of yeast and fungi (27). The
sample of MMO used in the present studies contained
an unknown impurity that was judged to be about
5% or less of the main product by gas-liquid-chroma-
tography. There was no detectable trace of the
hydrolysis product of MMO, hydroxymethyl-*myo*-inositol.

As seen in Fig. 9, pollen grains incubated in

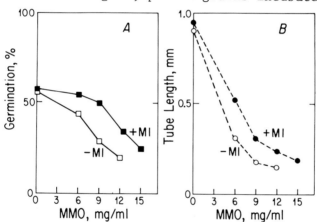

Fig. 9 Effect of 2-0,C-methylene-*myo*-inositol
on germination and tube elongation of L. *longiflorum*
cv. Ace pollen incubated in pentaerythritol medium
in the presence or absence of *myo*-inositol, 0.1 mg/ml.

standard medium in the presence of MMO for 5 hr produced shorter tubes and had a lower germination than
controls, increasingly so at higher levels of MMO.
In the absence of myoI in the medium, 6 mg per ml of

MMO reduced tube elongation by 60 to 70% while at the same concentration of MMO, germination was reduced only 20%. For this reason, 6 mg per ml of MMO was chosen as the concentration most useful in subsequent experiments. Addition of myoI to the MMO-containing medium, 100 µg per ml of myoI in the experiments described in Fig. 9, partially blocked inhibition by MMO.

A kinetic study of the effect of MMO on myoI conversion to tube wall pectin revealed competitive inhibition (Fig. 10). It is thought that MMO inhibits

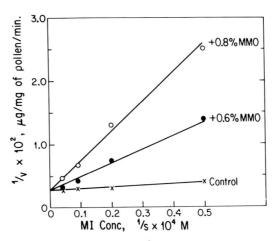

Fig. 10 Kinetics of 2-0,C-methylene-*myo*-inostiol inhibition of tritium incorporation into pollen tube wall polysaccharides in L. *longiflorum* cv. Ace pollen incubated in pentaerythritol medium containing *myo*-inositol-2-^{3}H.

oxidative conversion of myoI to D-glucuronic acid (28) but the enzymatic step affected by MMO in higher plants has not been clearly established.

Table 3 compares the conversion of *myo*-inositol-2-^{3}H and D-glucose-1-^{14}C to polysaccharides of pollen tube cytoplasm and wall in the presence and absence

Table 3

Effect of MMO on incorporation of myo-inositol-2-^{3}H and D-glucose-1-^{14}C

into lily pollen

Label	MMO	Cytoplasm	Wall
		%	%
myo-inositol-2-^{3}H	-	8.5	10.5
	+	4.4	2.7
D-glucose-1-^{14}C	-	32	34
	+	32	34

of MMO. Pollen grains were germinated and grown for
3 hr under standard conditions in regular media. Then
6 mg per ml of MMO was added to half of the samples.
The remaining samples were held as controls. Six
hr after germination, media of all samples was replaced with fresh solution containing either *myo*-inositol-
2-^{3}H or D-glucose-1-^{14}C and incubation continued for
3 hr, a total of 9 hr of growth. During the final 3
hr, 10.8% of the myoI was converted to wall polysaccharide by the control while the MMO-inhibited samples
incorporated only 2.7%, one-fourth of the control.
Conversion of label from D-glucose into wall polysaccharides was apparently unaffected by the presence of
MMO when judged on the basis of total incorporation
of label. Superficially, one might conclude from these
observations that the myoIOP played no part in pectic
polysaccharide biosynthesis when D-glucose was the
carbon source.

Quite a different story is revealed if saccharide
components of the tube wall are examined for distribution of label. Aliquots of the labeled wall fractions were hydrolyzed in 2 M trifluoroacetic acid
(121 C, 40 min, sealed tube). After passage of the

21

hydrolyzates through cationic and anionic exchange resins, the neutral sugar fractions were separated by paper chromatography. Chromatograms were scanned for radioactivity. These scans are reproduced in Fig. 11 (*myo*-inositol-2-^{3}H labeled pollen tube walls)

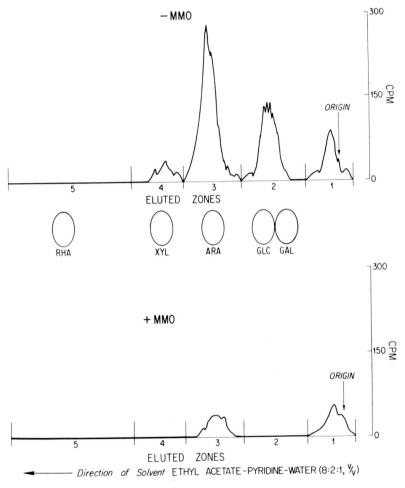

Fig. 11 Effect of 2-*O*,*C*-methylene-*myo*-inositol on the incorporation of tritium from *myo*-inositol-2-^{3}H into neutral sugar residues of tube wall polysaccharide of *L. longiflorum* pollen.

and Fig. 12 (D-glucose-1-^{14}C labeled pollen tube walls).

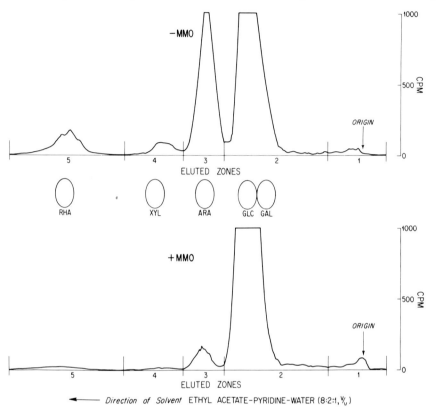

Fig. 12 Effect of 2-O,C-methylene-myo-inositol on the incorporation of ^{14}C from D-glucose-1-^{14}C into neutral sugar residues of tube wall polysaccharides of $L.$ $longiflorum$ pollen.

In Fig. 11, most of the tritium appeared in arabinose although some recycled into hexose as a consequence of the long growth period used in these experiments. MMO repressed myoI utilization to the point where only a small quantity of labeled arabinose was formed. This observation indicates that MMO does block the myoIOP somewhere beyond free myoI. When glucose was used to furnish the label (Fig. 12), labeled hexose accounted for a major portion of the radioactivity found in neutral sugar residues but an appreciable arabinose peak was also detected in the scan. MMO nearly abolished the arabinose peak but left the hexose peak intact. In fact, quantitation of counts revealed that most of the radioactivity missing from the arabinose peak due to MMO could be accounted for the hexose peak. Thus it appears that MMO does block the myoIOP quite specifically and that the portion of labeled glucose destined for uronic acids and pentoses via myoI is diverted when MMO is present, into other products consisting of hexose units.

A balance sheet of labeled saccharides recovered from acid hydrolyzates of tube wall polysaccharides after labeling with D-glucose-1-^{14}C is given in Table 4. MMO treated samples contained only about one-third

Table 4

Effect of MMO on the distribution of label from

D-glucose-1-^{14}C into tube wall carbohydrate

Component	Radioactivity	
	- MMO	+ MMO
	%	%
Uronic acid	7.5	2.5
Galactose + glucose	76	92
Arabinose	9	3
Xylose	1	0.5
Rhamnose	2	1
Total carbohydrate label recovered	95.5	99

as much ^{14}C in uronic acid and arabinose as the corresponding controls. As pointed out above, all of
this label shifted to hexose. The decrease in labeled rhamnose in MMO treated samples probably reflects
reduced biosynthesis of galacturonorhamnan (see
papers by G. O. Aspinall and by P. Albersheim et al.,
this volume). Even though rhamnose biosynthesis was
unaffected by MMO, its incorporation into wall polysaccharide was probably dependent upon galacturonic
acid biosynthesis.

These MMO studies provide the best evidence to
date that the myoIOP holds a vital role in the biosynthesis of pectic substance during pollen tube
wall formation. Our present experiments indicate
that at least 70% of uronic acid biosynthesis occurs
over this pathway. It should be pointed out that
the concentration of MMO used, 6 mg per ml, reduced
tube growth to about one-third of the control. This
residual growth may represent the remaining uninhi-
bited activity of the myoIOP and it may be that all
uronic acid formed is derived over this pathway.

Acknowledgements

The authors gratefully acknowledge the skillful
assistance of Miss Anne Golebiewski in these studies.
They also wish to thank Mr. James Stamos for help in
preparing illustrative material. Dr. Ramesh Shah
prepared the MMO used in this research.

References

1. F. Loewus, S. Kelly and E. F. Neufeld, *Proc.
    Natl. Acad. Sci. U.S. 48*, 421 (1962).
2. F. Loewus, and S. Kelly, *Biochem, Biophys, Res.
    Commun. 7*, 204 (1962).
3. L. Maquenne, *Les Sucres et Leurs Principaux
    Derives*, p. 189, Paris (1900).
4. F. Loewus, *Phytochemistry 2*, 109 (1963).

5. I-W. Chem and F. C. Charalampous, *J. Biol. Chem.* *239*, 1905 (1964); *241*, 2194 (1966).
6. L. Anderson in *The Carbohydrates, Chemistry/Biochemistry*, Ed. W. Pigman and D. Horton, Academic Press, N. Y. vol. IA, 1972, p. 519.
7. F. Eisenberg, Jr. in *Cyclitols and Phosphoinositides*, Ed. H. Kindl, Pergamon Press, N.Y., 1966, p. 3
8. M. W. Loewus and F. Loewus, *Plant Physiol.* *48*, 255 (1971).
9. I-W. Chem and F. C. Charlampous, *Arch. Biochem. Biophys.* *117*, 154 (1966); F. Eisenberg, Jr., *J. Biol. Chem.* *242*, 1375 (1967).
10. K. M. Gruhner and O. Hoffmann-Ostenhof, *Z. physiol. Chem.* *346*, 278 (1966); *Monatsh. Chem.* *99*, 1827 (1968).
11. F. Loewus, *Ann. N. Y. Acad. Sci.* *165*, 557 (1969).
12. D. S. Feingold, E. F. Neufeld and W. Z. Hassid, *Arch. Biochem. Biophys.* *78*, 401 (1958); E. F. Neufeld, D. S. Feingold and W. Z. Hassid, *Arch. Biochem. Biophys.* *83*, 9 (1959).
13. R. M. Roberts, *J. Biol. Chem.* *246*, 4995 (1971).
14. F. Eisenberg, Jr., A. H. Bolden and F. Loewus, *Biochem. Biophys. Res. Commun.* *14*, 419 (1964); I-W. Chen and F. C. Charalampous, *Biochem. Biophys. Res. Commun.* *17*, 521 (1964).
15. F. Loewus and S. Kelly, *Arch. Biochem. Biophys.* *102*, 96 (1963); F. Loewus, *Arch. Biochem. Biophys.* *105*, 590 (1964).
16. H. Kindl, see Ref. 7, 15; I-W. Chen and F. C. Charalampous, *Biochim. Biophys. Acta 136*, 568 (1967); G. Hauska and O. Hoffmann-Ostenhof, *Z. Physiol. Chem.* *348*, 1558 (1967); J. E. G. Barnett and D. L. Corina, *Biochem. J.* *108*, 125 (1968); F. Eisenberg, Jr. and A. H. Bolden, *Fed. Proc.* *27*, 595 (1968); E. Pina, Y. Saldana, A. Brunner and V. Chagoya, *Ann. N. Y. Acad. Sci.* *165*, 541 (1969).
17. W. R. Sherman, M. A. Stewart and M. Zinbo, *J. Biol. Chem.* *244*, 5703 (1969).

18. P. J. Harris and D. H. Northcote, *Biochem. J.*
    120, 479 (1970); P. H. Rubery and D. H.
    Northcote, see Ref. 25; C. Labarca, M. Kroh,
    and F. Loewus, *Plant Physiol.* *46*, 150 (1970);
    W. G. Rosen and H. R. Thomas, *Am. J. Bot.* *57*,
    1108 (1970); P. Jung, W. Tanner and K. Wolter,
    Phytochemistry *11*, 1655 (1972).
19. F. Loewus, *Fed. Proc.* *24*, 855 (1965).
20. M-S Chen, Ph.D. Dissertation, State University
    of New York at Buffalo, 1972.
21. I-W. Chen and F. C. Charalampous, *J. Biol.*
    Chem. *240*, 3507 (1965); H. Kurasawa, T. Hayakawa
    and S. Notoda, *Agr. Biol. Chem.* *31*, 382 (1967);
    V. H. Ruis, E. Molinari and O. Hoffmann-Ostenhof,
    Z. physiol. Chem. *348*, 1705 (1967); J. E. G.
    Barnett, R. E. Brice, and D. L. Corina, *Biochem.*
    J. *119*, 183 (1970); A. Brunner, L. M. Z. Pina,
    V. Chagoya de Sanchez and E. Pina, *Arch. Biochem.*
    Biophys. *150*, 32 (1972).
22. F. Loewus and Mary Loewus, *Plant Physiol.* *47*,
    S-5 (1971).
23. M. W. Loewus and F. Loewus, *Fed. Proc.* *31*, 882
    (1972).
24. M. W. Loewus and F. Loewus, *Plant Physiol.* In
    press (1973).
25. D. S. Feingold and J. S. Schutzback, *J. Biol.*
    Chem. *245*, 2476 (1970); D. T. A. Lamport,
    Annu. Rev. Plant Physiol. *21*, 235 (1970);
    P. H. Rubery and D. H. Northcote, *Biochim.*
    Biophys. Acta *222*, 95 (1970); A. A. Abdul-Baki
    and P. M. Ray, *Plant Physiol.* *47*, 537 (1971);
    J. E. Hopper, M. D. Davies and D. B. Dickinson,
    Plant Physiol. *47*, S-42 (1971).
26. M. Kroh and F. Loewus, *Science* *160*, 1352 (1968).
27. W. H. Schopfer and T. Posternak, *Z. Pathol. Bak-*
    teriol. *19*, 647 (1956); W. H. Schopfer,
    T. Posternak and D. Wustenfeld, *Arch. Mikrobiol.*
    44, 113 (1962).
28. P. A. Weinhold and L. Anderson, *Arch. Biochem.*
    Biophys. *122*, 529 (1967).

A STUDY OF POLLEN ENZYMES INVOLVED IN SUGAR NUCLEOTIDE FORMATION*

David B. Dickinson, James E. Hopper** and
Michael D. Davies***

Department of Horticulture, University of Illinios
Urbana, Illinois   61801

Abstract

Three enzymes (UDP-glucose pyrophosphorylase,
UDP-glucose dehydrogenase and glucuronokinase) from
pollen of Ace lilies, *Lilium longiflorum,* were partially purified and characterized. These enzymes
are located in pathways leading to sugar nucleotide
precursors of cell wall polysaccharides. The enzymes
are likely sites of *in vivo* feedback inhibition because each is inhibited *in vitro* by one or more sugar
nucleotides. UDP-glucose pyrophosphorylase (EC 2.7.
7.9) is inhibited by UDP-glucose, UDP-glucuronic acid,
UDP-galacturonic acid, UDP-xylose, and UDP-galactose.
These compounds had an additive effective effect when
present together. The sugar nucleotides exhibited

---

*Supported by NSF Grant GB-8764.

**Present address: Department of Biochemistry,
University of Washington, Seattle, Washington.

***Present address: Instituto de Biologia, Escola
Superior de Agricultura, Universidade Federal de
Vicosa, Vicosa, Minas Gerais, Brazil.

mixed competitive-noncompetitive inhibition with
the substrate UTP. In this respect the pollen enzyme differs from the calf liver enzyme. The
latter was also inhibited by the sugar nucleotides
listed above, but the inhibition was competitive
with UTP. UDP-glucose dehydrogenase (EC 1.1.1.22)
was inhibited by UDP-xylose, UDP-glucuronic acid
and UDP-galacturonic acid. Michaelis-Menten kinetics
were observed with the latter two compounds, but
sigmoid rate curves were observed with UDP-xylose.
Glucuronokinase (EC 2.7.1.43) was inhibited by
UDP-glucuronic acid and D-glucuronic acid-1-P.
These inhibitions may operate *in vivo* to link the
rates of sugar nucleotide production to their rates
of utilization for synthesis of cell wall polysaccharides.

Introduction

Metabolic pathways concerned with plant cell
wall polysaccharides are outlined in Fig. 1. Simple

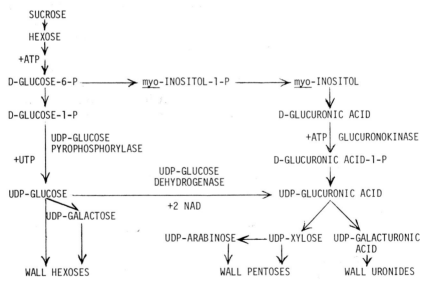

Fig. 1 Metabolic pathways from sugars to wall
polysaccharides of lily pollen.

sugars are the starting point, and the last water-
soluble intermediates are sugar nucleotides. The
latter are substrates for the various polysaccharide
synthetases although glycolipid intermediates may be
involved in some cases (1). A number of branch
points exist; the hexose phosphates may be incorporated into starch via ADP-glucose (2) or shunted
into the pentose phosphate and glycolytic pathways.
The UDP-glucose pyrophosphorylase reaction is an
important branch point because after conversion to
UDP-glucose the glucosyl residues would probably not
be available for glycolysis, the pentose phosphate
pathway, or reserve polysaccharides. Another major
branch point is at UDP-glucose dehydrogenase since
this step is the first in a series of reactions by
which cell wall pentoses and uronides may arise.
Inhibition of these enzymes by the sugar nucleotide
precursors of wall polysaccharides could be a regulatory mechanism whereby synthesis of sugar nucleotides
is linked to their rates of utilization. There is
some evidence that this is the case (3-5).

The inositol pathway (Fig. 1) provides a second
route for production of wall pentoses and uronides.
This pathway may be important in a number of plant
tissues (6, 7) including germinating lily pollen (8).
Sorghum seedlings apparently lack UDP-glucose dehydrogenase while at least one enzyme of the inositol
pathway (UDP-glucuronic acid pyrophosphorylase) is
present in abundance (9). The extensive studies of
Loewus and colleagues (6, 7) provide valuable information about the operation of the inositol pathway.
Labeled glucuronic acid accumulated but subsequent
intermediates of the inositol pathway did not accumulate when corn root tips were incubated with labeled
myo-inositol (7). This labeling pattern indicates
that phosphorylation of glucuronic acid was a rate-
limiting step, perhaps because of a relatively low
level of the phosphorylating enzyme or feedback
inhibition by end-products of the pathway. Glucuronokinase has been isolated from mung bean seedlings

31

(10), but there is no information concerning its
regulatory properties.

This paper is concerned with regulatory properties of three enzymes (UDP-glucose pyrophosphorylase, UDP-glucose dehydrogenase, glucuronokinase)
located in the pathways leading to cell wall polysaccharides. The enzymes were partially purified
and assayed in the presence of compounds which might
cause feedback inhibition *in vivo*. Lily pollen was
the source of enzyme because germinating lily pollen
is capable of rapid tube growth (11), and the pollen
tube wall resembles other primary cell walls (12).
Our earlier studies revealed that a number of enzymes
were abundant in mature lily pollen and that none
increased during germination (13). Therefore it is
likely that fluctuations in rates of tube wall
polysaccharide biosynthesis are accompanied by altered degrees of activation or inhibition of the enzymes
involved.

Materials and Methods

Ace lilies (*Lilium longiflorum*) were the source
of pollen. Pollen collection and storage procedures
were described earlier (11).

UDP-glucose pyrophosphorylase (EC 2.7.7.9).
This enzyme was isolated from nongerminated lily
pollen and purified 34-fold to a specific activity
of 32 μmoles glucose-1-P/min/mg protein as described
earlier (14). The enzyme from calf liver was purified by a procedure involving addition of protamine
sulfate to the crude extract, precipitation of the
enzyme between 40% and 60% ammonium sulfate and
column chromatography on DEAE-cellulose (J. Hopper
and D. Dickinson, manuscript in preparation). The
liver enzyme was purified 25-fold to a specific
activity of 9 μmoles UDP-glucose/min/mg protein.
The liver and pollen pyrophosphorylases were essentially free of contaminating enzymes capable of

hydrolyzing or modifying UTP, α-D-[^{14}C]glucose-1-P, UDP-D-[^{14}C]glucose, UDP-D-[^{14}C]glucuronic acid or UDP-D-[^{14}C]xylose. Kinetic studies of the purified enzymes were conducted with a radiochemical assay in which α-D-[^{14}C]glucose-1-P was converted to UDP-D-[^{14}C]glucose and the product was determined by scintillation counting (14). The back reaction was prevented by addition of inorganic pyrophosphatase to the reaction mixture.

UDP-glucose dehydrogenase (EC 1.1.1.22). UDP-glucose dehydrogenase was isolated as a soluble enzyme from pollen germinated 2 hr. This was done to avoid the detergent treatment which was required to detect enzyme activity in extracts of non-germinated pollen. Details of the procedures for enzyme purification and assay are given elsewhere (15). The enzyme was purified 12-fold to a specific activity of 1.8 nmoles UDP-glucuronic acid/min/mg protein. No contaminating enzymes capable of reacting with NADH, UDP-D-[^{14}C]glucose, UDP-D-[^{14}C]glucuronic acid, or UDP-D-[^{14}C]xylose were detected in the purified enzyme.

Glucuronokinase (EC 2.7.1.43). Chemicals included D-glucuronic acid (Grade 1), α-D-glucuronic acid-1-P (K salt), ATP (disodium salt), UDP-glucuronic acid (triammonium salt), and HEPES buffer from Sigma. Tween 80 was from Nutritional Biochemicals Corp., and ammonium sulfate (enzyme grade) was from Mann Research Lab. Miracloth and protamine sulfate (salmine) were from Calbiochem. *Escherichia coli* alkaline phosphatase (Type BAPSF) was from Worthington Biochemical Corp. DEAE-cellulose paper (DE-81 and DEAE-cellulose for column chromatography (DE-52, microgranular) were from W&R Balston, Ltd. Uniformly labeled D-[^{14}C]glucuronic acid, K salt, 76 Curies/mole, was from Amersham/Searle. The purity and identity of biochemicals were established using paper or ion-exchange paper chromatography. In particular, the absence of glucuronic acid from glucuronic acid-1-P and the absence of these two compounds from UDP-glucuronic acid were confirmed at the time of the enzyme inhibitor studies. The UDP-glucuronic acid also lacked chromatographically detectable UMP, and UDP was not detected using an enzymatic determination (16) capable of measuring less than 1% contamination by UDP. The ATP contained approximately 0.5% ADP according to enzymatic determination (16).

The enzyme was isolated by grinding pollen and isolation medium (100 mg pollen/ml medium) 4 min in an ice cold mortar. The medium (pH 7.5) contained 10 mM HEPES, 1mM EDTA, and 5 mM reduced glutathione. About 90% of the pollen grains were ruptured by this procedure. The homogenate was filtered through two layers of miracloth, and the filtrate was centrifuged (15 min, 25,000g, 0°C). The supernatant fluid was the source of enzyme. The standard reaction mixture (pH 7.2) contained pollen enzyme, 5 mM ATP, 10 mM $MgCl_2$, 1 mM D-$[^{14}C]$glucuronic acid (2,480 cpm/nmole) and 0.1 M HEPES in 50 µl final volume. The reaction was initiated with enzyme and terminated by 30 sec at 100°C after 10 min of incubation at 30°C. A 25 µl portion of each reaction mixture was placed on the origin of a 6 x 1¼ inch DEAE-paper strip and the strips were developed 2 or 2½ hr with 0.1 M ammonium acetate, pH 7.1. The strips were passed through a radiochromatogram scanner and the radioactive peaks corresponding to authentic α-D-glucuronic acid-1-P were cut out and placed in scintillation bottles. Samples were counted with a PPO–POPOP–toluene scintillation fluid as described earlier (14). One unit of enzyme activity = 1 µmole glucuronate phosphorylated/min.

Procedures used to characterize the labeled product of glucuronokinase included 1) chromatography on Whatman No. 1 paper with butanol:acetic acid:water (12:5:3), and 2) electrophoresis on Whatman 3 MM paper at 6½ to 7 volts/cm with 0.1 M ammonium formate as electrolyte and caffeine to mark the origin (10).

34

Results and Interpretation

An extensive search was conducted for feedback inhibitors of UDP-glucose pyrophosphorylase. Several sugar nucleotides were inhibitory to the pollen enzyme, and the inhibitor constants (K_i) are given in Table 1. Each sugar nucleotide gave mixed competitive-

TABLE 1

Inhibition of UDP-Glucose Pyrophosphorylase by Sugar Nucleotides:

Inhibitor Constants and Type of Inhibition

Source of enzyme	Inhibitor	K_i ,mM*	Type of inhibition
Lily pollen	UDP-glucose	0.13	Mixed competitive -
	UDP-glucuronate	0.75	noncompetitive with
	UDP-galacturonate	0.93	UTP and
	UDP-xylose	1.6	noncompetitive
	UDP-galactose	4.8	with glucose-1-P.
Calf liver	UDP-glucose	0.005	Competitive with
	UDP-glucuronate	0.15	UTP and non-com-
	UDP-galacturonate	0.21	petitive with
	UDP-xylose	0.24	glucose-1-P.

*UTP varied

non-competitive inhibition with the substrate UTP (V_{max} decreased, K_m increased) and non-competitive inhibition with glucose-1-P (V_{max} decreased, K_m unchanged). Of the various sugar nucleotide inhibitors, UDP-glucose had the lowest K_i and the greatest effect on V_{max} when UTP was varied. For instance, 4 mM UDP-glucose caused a 53% decrease in V_{max}, and 4 mM UDP-galacturonic acid caused a 32% decrease in V_{max}. UDP-glucose inhibits the same enzyme from rat liver

35

(17), erythrocyte (18), heart (18), and mung bean (18). However, these reports indicated that UDP-glucose inhibited competitively with respect to UTP (V_{max} unchanged, K_m increased). Accordingly, we purified and characterized a mammaliam UDP-glucose pyrophosphorylase to learn 1) whether it was inhibited by sugar nucleotides which inhibited the pollen enzyme, and 2) the type of inhibition. The bovine liver enzyme was studied by Hansen and colleagues (19), but there seems to be no information concerning its inhibition by UDP-glucose or other sugar nucelotides. Table I shows that the calf liver enzyme was indeed inhibited by several sugar nucleotides, and each inhibitor was competitive with UTP. UDP-glucose was by far the most effective inhibitor.

UDP-glucose pyrophosphorylase might be inhibited *in vivo* by several sugar nucelotides acting simultaneously. This possibility was explored by incubating the purified pollen enzyme with low concentrations of the inhibitors singly and in combination. Each compound apparently exerted a similar degree of inhibition whether alone or in the presence of other sugar nucleotides because an additive effect was observed (Table 2 and ref. 14).

UDP-glucose dehydrogenase. It was of considerable interest to find UDP-glucose dehydrogenase present in lily pollen although its activity in pollen extracts was less than 1% as great as the activity of UDP-glucose pyrophosphorylase (14, 15). In contrast, the final products of UDP-glucose dehydrogenase (wall pentose + uronide) account for at least 10% of the wall polysaccharide of lily pollen tubes (12, VanDerWoude, et al.). The significance of this is not clear. Perhaps UDP-glucose pyrophosphorylase acts *in vivo* at considerable less than maximal velocity. Perhaps significant amounts of wall pentoses and uronic acids are formed by a route which bypasses UDP-glucose dehydrogenase.

TABLE 2

Additive Inhibition of Pollen UDP-Glucose Pyrophosphorylase

by Sugar Nucleotides

Sugar nucleotides and concentrations	Assay 1	2	3	4	5	6	7	8
UDP-glucose 0.1 mM	+*				+	+	+	
UDP-xylose 0.1 mM		+			+	+		+
UDP-galacturonic acid 0.1 mM			+		+		+	+
UDP-glucuronic acid 0.1 mM				+	+			+
Observed % inhibition compared to a control lacking inhibitor**	12	4	8	8	32	17	19	24

*(+) Indicates presence of sugar nucleotides

**In the control lacking inhibitor, 7.0 nmoles

of UDP-glucose was formed. The substrates UTP

and glucose-1-P were held constant at 0.32 and

0.58 mM, respectively.

The kinetic constants of the purified pollen UDP-glucose dehydrogenase are summarized in Tables 3 and 4. The properties of this enzyme generally resemble those of pea cotyledon (3) and *Cryptococcus* (4) UDP-glucose dehydrogenase except that UDP-galacturonic acid was not shown to be an inhibitor of the latter two enzymes. UDP-glucuronic and UDP-galacturonic acids inhibited the pollen enzyme competitively with respect to the substrate UDP-glucose and noncompetitively with respect to the substrate NAD. All saturation curves were first order (hyperbolic). In constrast, UDP-xylose was a much more effective inhibitor and complex kinetics were observed. UDP-glucose saturation curves became sigmoid (slope of Hill plot approached 2) in the presence of UDP-xylose, and UDP-xylose inhibitor saturation curves were sigmoid when the substrate UDP-glucose was low.

37

TABLE 3

Pollen UDP-glucose Dehydrogenase

Summary of Substrate Constants

Substrate varied	S_{50} mM	n	Inhibitor added
1. UDP-glucose	0.3	1	No inhibitor
2. UDP-glucose	1.0	1	0.5 mM UDP-galacturonate
3. UDP-glucose	1.0	1	0.5 mM UDP-glucuronate
4. UDP-glucose	1.0	2	0.05 mM UDP-xylose
5. NAD	0.4	1	No inhibitor or with added UDP-glucuronate or UDP-galacturonate
6. NAD	0.8	1	0.05 mM UDP-xylose
7. NAD	1.5	1	0.10 mM UDP-xylose

TABLE 4

Pollen UDP-glucose Dehydrogenase

Summary of Inhibitor Constants

Inhibitor varied	K_i mM	I_{50} mM	n	Experimental conditions
1. UDP-xylose	0.016		1	Inhibitor saturation curves done at various levels of NAD, UDP-glucose held constant at 2 mM.
2. UDP-xylose		0.050	1	UDP-glucose at 0.75 mM, NAD at 4 mM.
3. UDP-xylose		0.008	2	UDP-glucose at 0.2 mM, NAD at 4 mM.
4. UDP-galacturonate	4.4		1	NAD varied as in No. 1 above.
5. UDP-galacturonate	0.46		1	UDP-glucose varied, NAD constant.
6. UDP-glucuronate	3.2		1	NAD varied, UDP-glucose constant.
7. UDP-glucuronate	0.40		1	UDP-glucose varied, NAD constant.

Sigmoid rate curves were not observed when UDP-glucose remained high and UDP-xylose was varied or when NAD was varied in the presence of inhibitory amounts of UDP-xylose.

It is clear from these results that the pollen enzyme is extremely complex and may have inhibitor binding sites distinct from the substrate site. It will be of interest to learn whether sigmoid rate curves are characteristic of all UDP-pentose inhibitors and first order curves are characteristic of UDP-hexose and UDP-uronic acid inhibitors.

Glucuronokinase. The standard glucuronokinase isolation and assay procedures (see Materials and Methods) were adopted after preliminary experiments with extracts of nongerminated lily pollen. ATP was the preferred nucleotide, and enzyme activity did not increase when pollen germinated 4 hr. When a detergent (5% v/v, Tween 80) and 0.5 M KCl were added to the standard isolation medium, no additional enzyme was found in the supernatant fluid, and insignificant enzyme activity (0.4% of that in the supernatant) was present in the resuspended pellet.

The partially purified enzyme (Table 5) was free of contaminating enzymes capable of hydrolyzing ATP, UDP-glucuronic acid, and glucuronic acid-1-P. These contaminating enzymes were assayed with approximately 10-times more purified pollen enzyme and incubation periods at least 3 times longer than were used for saturation curves and inhibitor studies. Formation of ADP from ATP was determined enzymatically (16); UDP-D-[^{14}C]glucuronate was used to check for contaminating phosphodiesterase; labeled glucuronic acid-P produced by the pollen enzyme (Fig. 3) was used to assay for contaminating phosphatase.

39

TABLE 5

Purification of Pollen Glucuronokinase

Fraction	Volume (ml)	Protein[1] (mg/ml)	Enzyme Activity		Purification (- fold)
			Units[2]	Units/mg protein	
1. Crude extract[3]	35.2	11.3	6.4 (100%)	0.016	1
2. Protamine[4]	34.7	9.2	5.9 (92%)	0.018	1.1
3. Ammonium sulfate[5]	2.7	41.6	4.3 (67%)	0.038	2.4
4. Sephadex G-200[6]	17.8	1.2	1.3 (21%)	0.060	3.8
5. DEAE-cellulose[7]	0.6	0.38	0.67 (10%)	0.90	56

[1] Determined according to Lowry, et al. (21).

[2] One enzyme unit = 1 μmole glucuronic acid phosphorylated/min. Calculations are based on the specific radioactivity of the [^{14}C] glucuronic acid substrate; uronic acid was determined according to Bitter and Muir (22).

[3] Prepared from 4.0 g pollen as described in Materials and Methods.

[4] Two percent protamine sulfate was added (0.06 ml/ml crude extract), and after 10 min the mixture was centrifuged (10 min, 25,000 xg, $0°$ C).

[5] The fraction precipitating between 30% and 45% saturated ammonium sulfate was resuspended in 5 mM KPO_4 - 2 mM dithiothreitol, pH 7.5.

[6] The equilibrating and eluting buffer was 5 mM KPO_4 - 2 mM dithiothreitol, pH 7.5.

[7] Enzyme was adsorbed, washed with several ml of 5 mM KPO_4 - 2 mM dithiotreitol, pH 7.5, and eluted with a linear gradient of KPO_4 (0.1 M to 1.0 M), pH 7.5 containing 2 mM dithiothreitol. The eluted enzyme was precipitated with 60% saturated ammonium sulfate, resuspended in 5 mM HEPES - 2 mM dithiothreitol, pH 7.5, and stored at $-23°$ C.

Most of the labeled glucuronic acid was converted to labeled product when an excess of purified enzyme was present in the reaction mixture (Fig. 2A and 3A). The labeled product had the same mobility as authentic α-D-glucuronic-1-P when subjected to paper and ion-exchange chromatography and paper electrophoresis. Incubation of the labeled product with alkaline phosphatase caused complete conversion to a compound which cochromatographed with authentic glucuronic acid. The K_m for glucuronic acid was about 0.3 mM (Fig. 4).

40

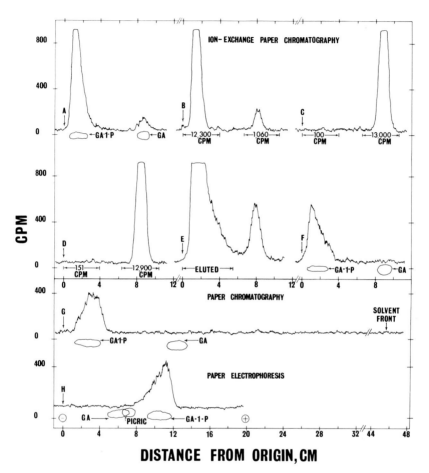

Fig. 2 Characterization of glucuronokinase reaction product. Purified pollen enzyme (3.8 μg protein) was incubated 60 min in the standard reaction mixture; 5 μl (A, B) and 25 μl (E) portions were chromatographed on DEAE-cellulose paper. Control reaction mixtures lacked ATP (C) or enzyme (D), and 5 μl was chromatographed. The counts/min listed under radioactive peaks were obtained by scintillation counting of the indicated segments of paper. Labeled product was eluted from E with 0.5 ml of 0.3 M ammonium acetate, pH 7.0. Portions (75 μl) of effluent were chromatographed on DEAE-
(legend continued on the following page)

41

cellulose paper (F) or Whatman No. 1 paper (G, 16 hr development). Trace H represents 120 μl of effluent after paper electrophoresis. See Materials and Methods for other details.

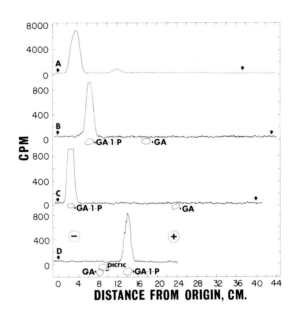

Fig. 3 Characterization of glucuronokinase reaction product. Sixty μl of reaction mixture, a replicate of the complete reaction mixture of Fig. 2, was chromatographed on paper (trace A, 24 hr development). The labeled product was eluted with 0.2 ml H_2O. Twenty μl portions of effluent were subjected to paper chromatography (B, 50 hr development), ion-exchange paper chromatography (C), and paper electrophoresis (D). Ammonium formate (0.1 M, pH 3.6) was the electrolyte used in paper electrophoresis.

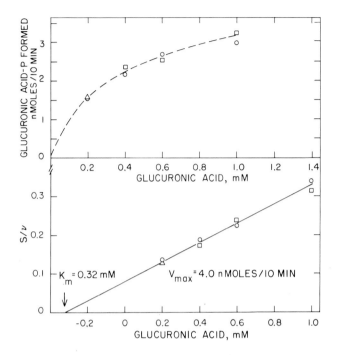

Fig. 4 Dependence of the velocity of purified pollen glucuronokinase on concentration of glucuronic acid (above), and a reciprocal plot of the same data (below). The standard assay procedure was used (see Materials and Methods) except that glucuronic acid was varied. The amount of enzyme added to various reaction mixtures was 0.19 μg (- △ -), 0.38 μg (- ○ -), or 0.76 μg (- □ -) protein, and data were normalized to 0.38 μg protein. In a plot of S/v vs S the X-intercept = $-K_m$ and the slope = $1/V_{max}$.

A search for inhibitors of glucuronokinase was conducted (Table 6). Most of the compounds tested were not inhibitory, but severe inhibition was caused by α-D-glucuronic acid-1-P and UDP-glucose. Further studies (Table 7) showed that greater than 50% inhibition occurred at 0.5 mM glucuronic acid-1-P

DAVID B. DICKINSON *et al.*

TABLE 6

Pollen Glucuronokinase: Inhibitor Survey

Percent inhibition	Compounds tested[*]
Less than 10%	Glucose, Fructose, Mannose, Galactose Galacturonic Acid, Gluconic acid Glucono-δ-Lactone, Myo-Inositol.
	Galactose-1-P, Glucose-1-P, Xylose-1-P Mannose-1-P, Fructose-1,6-P, Fructose-6-P
	UDP-Galacturonic Acid, UDP-Glucose UDP-N-Ac-Glucosamine, UDP-Xylose GDP-Mannose, UMP, UTP.
10 to 20%	UDP, GDP.
3Q to 50%	UDP-Mannose, UDP-Galactose.
Greater than 95%	Glucuronic Acid-1-P, UDP-Glucuronic Acid.

[*]All sugars and sugar acids were D and sugar-1-phosphates were α. Sugars and sugar acids were 10 mM in the standard reaction mixture: all other inhibitors were 4 mM in reaction mixtures containing 1 mM ATP.

TABLE 7

Pollen glucuronokinase: effect of various inhibitor concentrations

Glucuronic acid-1-P, mM	Inhibition (percent)	UDP-glucuronic acid, mM	Inhibition (percent)
0.5	64	0.4	42
1.0	74	0.8	58
2.0	91	2.0	90
4.0	100	4.0	99

44

and 0.8 mM UDP-glucuronic acid. The type of inhibition was not established, but it seems likely that the former compound, a product of the enzyme, is simply binding to the substrate site. The same may be true of UDP-glucuronic acid. Selectivity of the enzyme is indicated particularly by lack of inhibition from UDP-galacturonic acid and galacturonic acid. The enzyme probably does not catalyze phosphorylation of any of the sugars or sugar acids tested as inhibitors since these compounds in 10-fold excess over the substrate (glucuronic acid) did not diminish the rate of phosphorylation of glucuronic acid.

Discussion

Each of the three pollen enzymes studied was inhibited by one or more sugar nucleotides which may regulate activity of these enzymes *in vivo*. Each enzyme was inhibited by its reaction product so regulation could consist simply of a series of product inhibitions. However, additive inhibition by several sugar nucleotides, as shown *in vitro* for UDP-glucose pyrophosphorylase, could be advantageous to the cell. For instance, if small changes in steady state concentrations of several sugar nucelotides accompany altered growth rates, then the absolute change in total sugar nucleotide concentration would be greater than the change in a single sugar nucleotide, and a greater regulatory effect would be exerted on an enzyme which was sensitive to several sugar nucleotides rather than one. Also, additive inhibition by several metabolites would permit the pool size of each one to be smaller, decreasing the amount of uridine nucleotides required by the cell.

The mixed competitive-noncompetitive inhibition of the pollen UDP-glucose pyrophosphorylase may have regulatory significance. The noncompetitive component of the inhibition (effect on V_{max}) would tend to limit the ability of increased UTP to overcome inhibition of the enzyme by sugar nucleotides. Such

conditions may occur *in vivo* when the energy charge
is high, but cell wall synthesis is slowed, making
it desirable to inhibit the formation of sugar nucleotides. The properties of this pollen enzyme resemble
those of *E. coli* TDP-glucose pyrophosphorylase, the
first step in formation of TDP-rhamnose (20). TDP-
rhamnose exhibits a mixed type of inhibition, and
one possible interpretation of the effect is that
TDP-rhamnose binds to the substrate site as a competitive inhibitor and to a different site as a
noncompetitive inhibitor (20).

The great sensitivity of UDP-glucose dehydrogenase to inhibition by UDP-xylose may be relevant to
coordination of this enzyme with the inositol pathway.
When ample *myo*-inositol is available, such as during
breakdown of phytic acid, the flow of intermediates
through glucuronokinase might, in some plants, cause
sufficient accumulation of UDP-glucuronic acid and
then UDP-xylose to inhibit UDP-glucose dehydrogenase.
However, such a mechanism must remain speculative
until there is information concerning the relative
importance of the inositol pathway in a number of
plants and at various stages in their life cycles.
Furthermore, information is lacking as to the intracellular concentrations of the metabolites which
inhibited the pollen enzymes.

Acknowledgment

The authors are grateful to Dr. Frank Loewus for
advice and for a generous gift of glucuronic acid-1-P.

References

1. C. L. Villemez and A. F. Clark, *Biochem. Biophys.
Res. Commun. 36*, 57 (1969); H. Kauss, *FEBS LETT.
5*, 81 (1969).
2. J. Preiss and T. Kosuge, *Ann. Rev. Plant
Physiol. 21*, 433 (1970).

3. E. F. Neufeld and C. W. Hall, *Biochem. Biophys. Res. Commun.* 19, 456 (1965).
4. H. Ankel, E. Ankel and D. S. Feingold, *Biochemistry* 5, 1864 (1966).
5. A. A. Abdul-Baki and P. M. Ray, *Plant Physiol.* 47, 537 (1971).
6. F. Loewus, *Ann. Rev. Plant Physiol.* 22, 337 (1971).
7. F. Loewus, *Ann. N. Y. Acad. Sci.* 165 (2), 577 (1969).
8. M. Kroh and F. Loewus, *Science* 160, 1352 (1968).
9. R. M. Roberts, *J. Biol. Chem.* 246, 4995 (1971).
10. E. F. Neufeld, D. S. Feingold, W. Z. Hassid, *Arch. Biochem. Biophys.* 83, 96 (1959); D. S. Feingold, E. F. Neufeld and W. Z. Hassid in *Modern Methods of Plant Analysis* (H. F. Linskens, B. D. Sanwal and M. V. Tracey, eds), Vol. VII, Springer-Verlag, Berlin (1964) p. 474.
11. D. B. Dickinson, *Science* 150, 1818 (1965); D. B. Dickinson and D. Cochran, *Plant Physiol.* 43, 411 (1968).
12. W. V. Dashek, H. I. Harwood and W. G. Rosen in *Pollen: Development and Physiology* (J. Heslop-Harrison, ed.) Butterworths, London, (1971) p. 194; W. J. VanDerwoude, D. J. Morré and C. E. Bracker, *J. Cell Sci.* 8, 331 (1971).
13. D. B. Dickinson and M. D. Davies in *Pollen: Development and Physiology* (J. Heslop-Harrison, ed.) Butterworths, London (1971) p. 190; D. B. Dickinson and M. D. Davies, *Plant Cell Physiol.* 12, 157 (1971).
14. J. E. Hopper and D. B. Dickinson, *Arch. Biochem. Biophys.* 148, 523 (1972).
15. M. D. Davies and D. B. Dickinson, *Anal. Biochem.* 47, 209 (1972; M. D. Davies and D. B. Dickinson, *Arch. Biochem. Biophys.* 152, 53 (1972).
16. H. Adam, in *Methods of Enzymatic Analysis* (H. U. Bergmeyer, ed.) Academic Press, New York (1963) p. 573.

17. S. Kornfeld, *Fed. Proc.* 24, 536 (1965).
18. K. K. Tsuboi, K. Fukunaga and J. C. Petricciani, *J. Biol. Chem.* 244, 1008 (1969).
19. G. J. Albrecht, S. T. Bass, L. L. Seifert, and R. G. Hansen, *J. Biol. Chem.* 241, 2968 (1966); S. Levine, T. A. Gillett, E. Hageman, and R. G. Hansen, *J. Biol. Chem.* 244, 5729 (1969); T. A. Gillett, S. Levine and R. G. Hansen, *J. Biol. Chem.* 246, 2551 (1971).
20. R. L. Bernstein and P. W. Robbins, *J. Biol. Chem.* 240, 391 (1965).
21. O. H. Lowry, N. J. Rosebrough, A. L. Farr and R. J. Randall, *J. Biol. Chem.* 193, 265 (1951).
22. T. Bitter and H. M. Muir, *Anal. Biochem.* 4, 330 (1962).

UDP-D-GLUCURONIC ACID PYROPHOSPHORYLASE AND THE FORMATION OF UDP-D-GLUCURONIC ACID IN PLANTS

R. M. Roberts and J. J. Cetorelli

Department of Biochemistry
J. Hillis Miller Health Center
University of Florida
Gainesville, Florida   32601

Abstract

The enzyme UDP-D-glucuronic acid pyrophosphory-lase is widespread in plants and occurs in particularly high amounts in young, rapidly growing tissues such as germinating seeds and cells in culture. The enzyme has been partially purified from corn and barley and some of its characteristics examined. In seedlings, the enzyme is confined largely to the embryo and there is a close correlation between its levels and the rate of growth of the tissues. By contrast, UDP-D-glucose dehydrogenase appears to be more restricted in its distribution. However, during development of a sapwood cell from cambium in *Acer rubra*, in which there is a build-up of a thickened secondary wall, the activity of the dehydrogenase increases several fold, while the activity of the pyrophosphorylase declines. It is likely that in young primary tissues, UDP-D-glucuronic acid is provided via a pathway which involves *myo*-inositol as an intermediate and whose final step involves uridylyl transfer from UTP to D-glucuronic acid 1-phosphate. During the formation of the

secondary cell wall of *Acer rubra*, however, UDP-glucuronic acid is probably provided by the oxidation of
UDP-glucose.

Symbols

UDP-GlcUA = UDP-D-Glucuronic acid
DEAE   = diethylaminoethyl

Introduction

UDP-D-Glucuronic acid (UDP-GLcUA) is a crucial
intermediate in the formation of cell-wall polysaccharides in plants. Not only is it a glycosyl
donor itself, it is also the parent compound of
the related nucleoside diphosphate sugars UDP-D-xylose,
UDP-L-arabinose, UDP-D-apiose, and UDP-D-galacturonic
acid, which are themselves precursors of polysaccharides (1). As Dr. Loewus has already pointed out,
UDP-GlcUA can be formed in two ways in plant tissues:
a) from UDP-D-glucose by UDP-glucose dehydrogenase
(UDP-D-glucose:NAD$^+$ oxideroreductase E.C.1.1.1.22)
(2) and b) from D-glucuronic acid 1-phosphate by nuc-
leotidyl transfer from UTP (UTP-α-D-glucuronic acid
1-phosphate uridylyl transferase, class E.C.2.7.7.)
(3). Reaction (a) involves the oxidation of UDP-glucose by NAD$^+$. The enzyme was purified by Strominger
and Mapson from pea seedlings. (2) and resembled the
enzyme from liver in many of its properties (4).
Little information is available, however, about its
general distribution in plants or about its quantitative importance to the formation of UDP-GlcUA. An
enzyme catalyzing reaction (b) (UDP-GlcUA pyrophosphoyrlase) was first described in an ammonium sulfate
fraction prepared from mung beans (3), but has since
received little attention, possibly because it seemed
an enzyme of little significance except for the salvage of D-glucuronic acid produced during polysaccharide breakdown.

In 1962, however, Loewus and his colleagues (5)
described the metabolism of *myo*-inositol-2-^{14}C to

cell-wall uronosyl and pentosyl units in strawberry and parsley leaves. As we have seen earlier, this observation has been extended to a wide number of plant tissues, including barley and corn seedling tissue (6), and the available evidence suggests that myo-inositol, which is a ubiquitous component of plant tissues, may have a primary role as an intermediate in the biosynthesis of cell-wall polysaccharides. The pattern of labelling in the cell wall was similar to that observed with D-glucuronic acid-^{14}C and has indicated that myo-inositol-2-^{14}C is oxidatively cleaved in a sterospecific reaction to form D-glucuronic acid-5-^{14}C. It is the subsequent metabolism of the latter compound that accounts for the labelling of polysaccharide in the cell wall.

There is, therefore, a considerable amount of circumstantial evidence to implicate a pathway which involves the intermediate formation of myo-inositol and the terminal enzyme UDP-glucuronic acid pyrophosphorylase as a major source of UDP-GlcUA and hence, cell-wall material in a number of plant tissues. It is not easy, however, to differentiate quantitatively between two closely related biochemical pathways, both of which can be considered to begin with the same starting substrate, namely, D-glucose, and both of which give identical radioactive-labelling patterns in the final product. However, one criterion that has to be met in judging the importance of any bio-chemical route is whether the proposed enzymes are present in amounts potentially sufficient to account for the required rate of catalysis. From this point of view, therefore, we have investigated the distribution and activity levels of the enzymes UDP-GlcUA pyrophosphorylase and UDP-glucose dehydrogenase in a number of plant tissues.

As these enzymes are responsible for the final steps of the two alternative routes leading to UDP-GlcUA formation, we felt that such a study might reflect which pathway was operative or of primary

51

importance in the tissues. We also describe the
partial purification of UDP-GlcUA pyrophosphoyrlase
from germinating barley and corn and discuss some
of the properties of this enzyme.

Assays of the Enzymes

UDP-Glucuronic pyrophosphorylase and UDP-glucose
dehydrogenase are relatively unstable enzymes (7).
Moreover, their assay presents certain difficulties
which should be borne in mind in view of the results
presented in this paper. Therefore, we have devoted
one section to a consideration of the methods employed for measuring the activities of the two
enzymes.

Assay for UDP-GlcUA pyrophosphorylase. This enzyme cannot be assayed in crude tissue extracts,
even after dialysis, because of interfering
side reactions which lead to rapid hydrolysis of the
atarting substrates. We have found that it can be
assayed, however, in the protein fraction precipitated between 0 and 70% saturated ammonium sulfate.
In all these experiments, therefore, we have employed either this fraction or a narrower (35-60%)
cut in which more than 90% of the enzyme is collected
in order to estimate the levels of the pyrophosphorylase. The precipitate was redissolved in buffer and
dialyzed for 2 hours against 0.01 M Tris-HCl (pH 8.2)
before assay. The enzyme was assayed radiochemically
(7). The composition of the incubation mixture for
the forward reaction was as follows: $MgCl_2$ (1 μmole),
UTP ((1 μmole), GlcUA 1-P-U-^{14}C (0.25 μmole; 0.25 μCi),
Tris-HCl, pH 8.0 (10 μmole), enzyme (20 μl) in a final
volume of 0.1 ml; and for the reverse direction:
$MgCl2$ (1 μmole), sodium pyrophosphate (1 μmole), UDP-
GlcUA-U-^{14}C (0.2 μmole; 0.25 μCi), Tris-HCl, pH 8.2
(10 μmole), enzyme (20 μl) in a final volume of 0.1
ml. Incubation temperature was 30°, and 1 unit of
enzyme is defined as the activity required to produce
1 nmole of product per min at 30°.

Aliquots (5 μl) of the reaction mixtures were removed at times 0, 5, 10, 15, and 30 min in order to obtain initial rates of reaction, and were spotted on Whatman 1 paper. UDP-GlcUA was separated from the monophosphate by descending chromatography in ethanol-1 M ammonium acetate, pH 3.8 (5:2 v/v). Radioactive areas were detected by autoradiography, cut out, and counted by liquid scintillation. Radioactivity was expressed as a percentage of the total ^{14}C in product and substrate radioactivity together.

Assay for UDP-glucose dehydrogenase. This assay was based on that described by Strominger and Mapson (2). NAD^+ reduction was measured spectrophotometrically at 30⁰ and 340 nm, using a Gilford Model 222 recording spectrophotometer. The usual assay systems contained Tris-HCl, pH 8.2 (20 μmoles), NAD^+ (1 μmole), enzyme solution (variable) and water to make a final volume of 0.95 ml. The reaction was started by addition of UDP-glucose (0.05 ml; 0.5 μmole) and followed for at least 30 min. A blank without UDP-glucose was always run to obtain the rate of reduction of NAD^+ in absence of added substrate. A unit of activity is again defined as the amount of enzyme required to produce 1 nmole of UDP-GlcUA/min under the conditions of assay.

However, the minimum quantity of enzyme that can be measured in the plant extracts is limited by the rate of reoxidation of the NADH in the cuvette by contaminating NADH dehydrogenase activities, which are particularly abundant in plants. Moreover, the substrate UDP-glucose is again rapidly broken down in crude plant extracts. Therefore, we have usually assayed the enzyme in an ammonium sulfate precipitate (most frequently, that precipitated between 0 and 70% saturation), although we have also tested for its presence in the initial extract. The contribution of NADH dehydrogenase activity was assessed by measuring the activity of bovine UDP-glucose dehydrogenase (of known specific activity from Sigma) in presence of the enzyme preparation from the plants. The amount of beef enzyme required to give a measureable initial rate of NAD^+ reduction was assumed to represent the minimal amount of plant enzyme that could be measured reliably under the conditions of assay.

Because some of the enzymes related to nucleoside diphosphate sugar metabolism in plants have been reported to be associated with membranes, we have also assayed for UDP-glucose dehydrogenase activity in 0.1% v/v Tween extracts of the particulate fraction of our homogenates. In most instances, activities were very low or absent in this fraction, even when appreciable enzyme levels were detected in the initial buffer extracts.

Distribution of the Enzymes in a Number of Plant Tissues

Table I lists the activities of UDP-GlcUA pyrophosphorylase and UDP-glucose dehydrogenase in a number of plant tissues that we have tested. We realize that this group of plants is only narrowly representative. Nevertheless, it is clear that while the pyrophosphorylase appears to be widely distributed in higher plants, the dehydrogenase is found in measurable amounts in only a few species. It should be pointed out that the levels of pyrophosphorylase in most of these tissues are well in excess of what one would estimate is likely to be required for UDP-GlcUA in cell-wall polysaccharide biosynthesis. For example, 100 g of tissues from 5-day-old corn seedlings are potentially capable of synthesizing at least 143 μmoles of UDP-GlcUA/min. If the enzyme was operating at full capacity *in vivo* and all of the product was converted to cell-wall material, approximately 35 g of the new polysaccharide could be formed per day. This is almost two orders of magnitude more than would be required for normal growth of the seedling.

TABLE 1

THE ACTIVITIES OF UDP-D-GLUCURONIC ACID PYROPHOSPHORYLASE AND
UDP-D-GLUCOSE DEHYDROGENASE IN A RANGE
OF PLANT TISSUES

Plant tissue	Units of enzyme activity (μmoles/min/100 g tissue)	
	Pyrophosphorylase	Dehydrogenase
Zea mays		
(ungerminated grain)	40	ND
(5-day-old seedling)	143	ND
Hordeum vulgare		
(ungerminated grain)	45	ND
(5-day-old seedling)	50	ND
Triticum vulgare	15	--
(ungerminated grain)		
Vicia faba		
(3-day-old seedling)	40	--
Lemna gibba		
(whole plant)	41	ND
Wolffiella floridana		
(whole plant)	ND	0.15
Dauca carota		
(washed slices)	10	--
Cucurbita pepo (fruit)	0.38	ND
Brassica sp.		
(flower buds)	0.73	ND
Pisum sativum		
(immature pea in pod)	20	0.31
(3-day-old seedling)	21	0.56

The symbol ND means that activity was not detectable under the conditions
of the assay. In certain instances the dehydrogenase was not assayed (--).

The absence of pyrophosphorylase in *Wolffiella floridana* is particularly significant, as this plant is also unusual in that it does not metabolize *myo*-inositol to cell-wall polysaccharide (8). Rather, it converts it to *myo*-inositol phosphates of various kinds. Presumably, in this tissue the UDP-GlcUA is synthesized from UDP-glucose, as the dehydrogenase is present in significant amounts. Even in this instance, the dehydrogenase could provide precursor for more than 40 mg of new polysaccharide in one day, much more than would be required by this relatively slow-growing plant. *Wolffiella* is clearly an exception, however, to the general rule that primary tissues contain high levels of pyrophosphorylase activity.

In peas, both enzymes can be detected. It has not yet been established, however, whether the two are localized in separate tissues, or how their relative activities might change with the growth and development of the young seedlings or the maturing seed.

UDP-GlcUA Pyrophosphorylase Activity During Germination of Barley and Corn

Seedling tissues provide particularly good material for studies on enzymes involved in sugar nucleotide metabolism because of their rapid growth and high rate of cell-wall synthesis. In barley, the activity of the pyrophosphorylase was determined in both underminated grain and in seedlings germinated sterilely on 0.5% agar for 6 days.

In the experiment reported in Fig. 1, the average fresh weight of each seedling increased from about 38 mg at day 0 to around 200 mg at day 5. Most of this increase is due to the growth of the embryo. The enzyme activity also increased markedly during

56

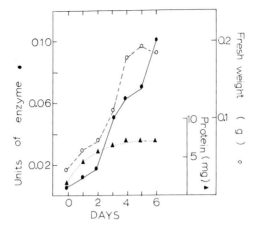

Fig. 1   Increase in UDP-glucuronic acid pyrophosphorylase activity during germination of barley grain.
Total enzyme activity per plant (●), average weight
of each seedling (O), and total protein in the enzyme
preparation (▲) were followed at daily intervals.   A
unit of activity is defined as the enzyme required to
produce 1 µmole GlcUA 1-P per min in the reverse reaction.   Assayed in the forward direction, the rates
of reaction and, hence, number of units would be
about 1.25 times that in the reverse.

germination, and the total number of units extracted
per seedling increased more than 15-fold during the
experimental period.   By day 6, each young plant contained sufficient enzyme to produce more than 6 µmole
of UDP-GlcUA per hour.

A similar series of experiments was conducted
with corn (Fig. 2), except the embryos in these instances were separated from the rest of the seed
before extraction of the enzyme.   Note that, whereas
the activity of UDP-GlcUA pyrophosphorylase remained
relatively constant in the endosperm, there was a
very rapid increase in the embryo which coincided
very closely with the growth of the young plant.   The

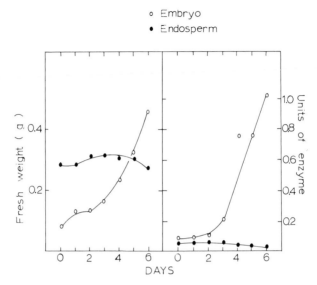

Fig. 2 UDP-Glucuronic pyrophosphorylase activity in the embryo (O) and endosperm (●) of corn.
One unit of enzyme activity represents a rate of reaction of 1 μmole per min in the forward reaction.
In each experiment, groups of about 50 seedlings were
used and the results then calculated for single seedlings. Changes in embryonic and endospermic fresh
weight are shown on the left.

increase in embryonic fresh weight noted during day
1 seems to be due to initial hydration of the tissues,
and it is only after about 36 hours that the radicle
emerges from the grain. Growth proceeds very rapidly after day 2, and only then do the enzyme levels
start to increase. It should be noted that the
total amount of protein per seedling remains approximately constant during these experiments, as
the plants are germinated on agar, and have to rely
entirely upon their own nitrogen reserves.

At no stage in the germination of barley or corn
have we been able to detect any UDP-glucose dehydro-

genase activity, even under condition in which the
bovine enzyme is fully active in presence of plant
extract. Other buffers, changes in the concentration of EDTA and mercaptoethanol, and changes in
the composition of the assay mixture itself have
all been tried, but in no case have measurable rates
of reduction been detected. We conclude, therefore,
that it is the pyrophosphorylase which is most
likely to be responsible for UDP-GlcUA formation in
these rapidly growing seedling tissues. In particular, the close correlation between the activity of
this enzyme and the rate of growth of the young
plant is strong suggestive evidence for such a
role.

UDP-GlcUA Pyrophosphorylase and UDP-Glucose
Pyrophosphorylase along the Root Tip of *Zea mays*

Root tips of monocotyledons provide a very
useful system for studying some of the biochemical
changes which accompany growth and development,
because all cells originate in the meristem, which
is confined to the tip region. Segments of tissue
cut from the axis of such roots, therefore, provide
a series of cell populations of increasing developmental age. We have, therefore, investigated the
changes in UDP-GlcUA pyrophosphorylase that occur
along the primary root tip of a 3-day old corn seedlings. The initial 0.5 cm tip region from approximately 900 seedlings was cut into 5 one-mm segments
using a guillotine of razor blades separated by
one-mm spacers. The next 1.5 cm of the axis of each
root was cut into three 0.5-cm pieces. Each batch
of tissues was weighed and UDP-GlcUA pyrophosphorylase and UDP-glucose pyrophosphorylase (9) activities measured in the ammonium sulfate fraction
precipitated between 20 and 70% saturation. A time
course was run for each reaction in order to determine initial rates. This is particularly important
for UDP-glucose pyrophosphorylase, as this enzyme

shows pronounced product inhibition and the reaction rate rapidly slows. From Fig. 3 it is clear that the activity of UDP-GlcUA pyrophosphorylase is highest in segments 1 and 2. The amount or protein, the number of cells, and the amount of cell-wall material are also highest in this region (10, 11). Growth in cell length continues to about 6 mm and then stops (10). However, it is clear from Fig. 3 that the enzyme levels remain relatively high, even in the nongrowing regions (1.0 cm and beyond), presumably in order to provide precursors for wall consolidation and thickening. The level of UDP-glucose pyrophosphorylase was about ten times as high as UDP-GlcUA pyrophosphorylase throughout this 2-cm region of the root tip. This enzyme presumable provides all of the UDP-glucose required for cell-wall glucan synthesis, as GDP-glucose pyrophosphorylase was not detectable. Again, these results emphasize that both UDP-glucose and UDP-GlcUA pyrophosphorylase are present in tissues where rapid growth and active cell-wall synthesis are occurring.

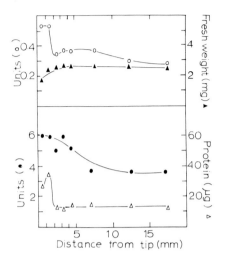

Fig. 3 Changes in UDP-glucuronic acid and pyrophosphorylase (0) and UDP-glucose pyrosphorylase (●)) along the root tip of corn (*Zea mays*). One unit of activity represents a rate of 1 nmole per minute. Fresh weight (▲) and enzyme protein (△) **per** mm segment are also shown.

Purification and Properties of UDP-GlcUA Pyrophosphorylase from Barley and Corn Seedlings

UDP-GlcUA has been partially purified from barley seedlings by precipitation with ammonium sulfate and chromatography on DEAE-cellulose (see 7). During the latter step it is eluted ahead of and is clearly separated from UDP-glucose pyrophosphorylase. A similar pattern of elution is noted with the enzyme from young corn seedlings.

The enzyme has a pH optimum between 8 and 9, whether assayed in the forward to the reverse direction. The equilibrium constant of the reaction

$$\text{UDP-GlcUA} + \text{PP}_i \overset{\rightarrow}{\leftarrow} \text{GlcUA 1-P} + \text{UTP}$$

at pH 8.2, 30°, and at equimolar concentrations of

61

Mg^{2+} and pyrophosphate is approximately 2.7. The partially purified enzyme has Km values of 0.33 mM for GlcUA 1-P in the forward direction and 0.5 mM for UDP-GlcUA in the reverse. It showed normal Michaelis-Menten kinetics and a straight-line relationship between $1/v$ and $1/s$. The V_{max} was 1.25 times greater in the forward than in the back reaction. There was no detectable activity when GTP, ITP, ATP, or TTP was substituted for UTP in the forward reaction. Like other pyrophosphorylases, the enzyme requires divalent cations for activity. Magnesium, manganese, cobalt, zinc, and calcium were found to be effective, in decreasing order. The enzyme activation by magnesium was examined at three different concentrations of UTP (7). The velocity-Mg^{2+} concentration curves were not noticeable sigmoid-shaped, and maximum rates were always found at Mg^{2+} concentrations approximately twice that of the UTP. Higher concentrations were somewhat inhibitory. Noncomplexed free UTP may also inhibit, as indicated by the lower velocities with 10 mM UTP than with 2.5 mM UTP at Mg^{2+} concentrations less than 10 mM. Little effect on the rate or on the Mg^{2+} dependency was observed by varying the GlcUA 1-phosphate concentration in relation to Mg^{2+}. This suggests that the effect of Mg^{2+} on the reaction rate is only in relation to UTP.

In all of these respects, the enzyme generally resembles most other pyrophosphorylases that have been examined (12, 13, 14 for example). The UTP:Mg^{2+} dependency of 1:2 is somewhat unusual, as most workers report a 1:1 relationship between nucleoside triphosphates and Mg^{2+}. However, we have seen a similar dependency during assay of UDP-glucose and ADP-glucose pyrophosphorylase from wheat (9). It is possible, therefore, that Mg_2UTP is the substrate for the enzyme.

We have also investigated the possibility that

62

UDP-GlcUA pyrophosphorylase might be a regulatory enzyme. However, no effects of 2 mM concentrations of glucose, glucose 6-phosphate, glucose 1-phosphate, *myo*-inositol, or GlcUA have been noted when these compounds were tested as possible activators of the enzyme. Similarly, UDP-xylose, UDP-glucose and UDP-galacturonic acid (each up to 1 mM) had no inhibitory effect on the initial rate of the forward reaction under standard assay conditions. Moreover, unlike UDP-glucose pyrophosphorylase, the enzyme is not subject to strong feedback control by its nucleoside dephosphate sugar product, UDP-GlcUA. This evidence leads us to suspect that the pyrophosphorylase is not a regulatory enzyme for the *myo*-inositol pathway.

UDP-GlcUA Pyrophosphorylase and UDP-Glucose Dehydrogenase Activity during Cambial Development

During development of sapwood from the meristematic cambial tissues of trees, there is a build up of a thickened secondary wall, and an almost tenfold increase in the weight of each cell (15). This transition occurs very rapidly as functional vessels can be found within five or six cell widths of the cambium (16). In hardwoods, the secondary wall of sapwood is comprised largely of lignin, α-cellulose, and acidic xylans (17). The latter consist of long, uninterrupted chains of $\beta1{\rightarrow}4$-linked xylose units, approximately one in ten of which is 2-O-substituted with a residue of 4-O-methyl D-GlcUA (17, 18). The massive synthesis of the xylose and uronic acid must require the intermediate formation of UDP-D-GlcUA. Rubery (19) has shown the specific activity of UDP-glucose dehydrogenase increases markedly as the cambial cells of *Acer pseudoplatanus* differentiate into sapwood. He suggested, moreover, that it is this enzyme and not the pyrophosphorylase that is responsible for UDP-GlcUA synthesis during the transition from primary to secondary growth. He did not, however, measure the activity of the latter

63

enzyme in either tissue, and so his hypothesis was largely speculative.

We have repeated Rubery's experiments using *Acer rubra*, which was stripped from three 7-year old trees and the cambial cells gently scraped away from the wood beneath, using a scalpel. Xylem from the ring of sapwood immediately below this cambial layer was sliced off using a sharp knife. Each tissue was ground up in buffer and the enzyme protein precipitated by addition of solid ammonium sulfate to 70% saturation. For comparative purposes, the activities of UDP-glucose pyrophosphorylase and GDP-glucose pyrophosphorylase and UDP-glucose dehydrogenase, were measured in these protein fractions.

It is clear from Table 2 that, although the activity of UDP-glucose dehydrogenase per 100 g tissue weight is slightly less in sapwood than in cambium as expressed on a total activity basis, the xylem cells possess the greatest specific activity of the enzyme. Further, as the average weight of each cambial cell of *Acer* has been calculated to be only 15 mμg (15), the activity of UDP-glucose dehydrogenase must increase approximately eightfold during the developmental transition of a cambial cell into sapwood. These results are very similar to those published by Rubery and indicate that the two maples *Acer pseudoplatanus* and *Acer rubra* provide directly comparable tissue.

TABLE 2

THE ACTIVITIES OF UDP-D-GLUCOSE PYROPHOSPHORYLASE,

UDP-D-GLUCURONIC ACID PYROPHOSPHORYLASE, AND UDP-D-GLUCOSE DEHYDROGENASE

IN TISSUES OF MAPLE (ACER SP.)

Tissue	Enzyme activity (nmoles/min/100 g tissue)		
	UDP-pyrophosphorylase	UDP-GlcUA pyrophosphorylase*	UDP-G dehydrogenase*
Callus** (A. pseudoplatanus)	14,000	1,800 (15)	140 (1.2)
Cambium (A. rubra)	24,700	370 (2)	180 (0.9)
Sapwood (A. rubra)	8,200	ND (-)	140 (4.2)

*Figures in parentheses refer to the specific activity of the enzyme in the enzyme preparation (nmoles/min/mg protein).

**This was found to have a variable UDP-GlcUA activity, depending upon its age. Young, rapidly growing tissue contained high levels; stationary cultures, little or no activity. The tissue here was harvested 30 days after transfer and was in an intermediate condition.

By contrast to the increase in the activity of the dehydrogenase, the level of UDP-GlcUA pyrophosphorylase declined markedly during wood formation. Indeed, we were unable to detect its presence in the preparations from sapwood, though the slightly variable blank values obtained in the radiochemical assay preclude our measuring activities of much less than 50 nmoles/min/100g of tissue. GDP-glucose pyrophosphorylase was undetectable in both the cambium and the sapwood. However, the ubiquitous UDP-glucose pyrophosphoyrlase was present in both tissues in high amounts and is presumable responsible for the formation of the sugar nucleotide precursor of cellulose.

Our results, therefore, suggest strongly that the direct oxidation of UDP-glucose provides the UDP-GlcUA requires for secondary cell-wall formation in xylem tissue of *Acer*, whereas in younger tissues the high levels of the pyrophosphorylase indicate that it is the pathway from *myo*-inositol to UDP-GlcUA that predominates. The results with callus cells (obtained from *Acer psuedoplatanus*), which have only primary cell walls, tend to confirm this impression (Table II).

Summary and Conclusions

The activity levels of the two enzymes UDP-GlcUA pyrophosphorylase and UDP-glucose dehydrogenase have been investigated in a variety of plant tissues. Although it is recognized that an inability to detect an enzyme is not proof of its absence and that lack of suitable activators, the presence of inhibitors, or extreme instability of the enzyme protein could account for apparent low activity levels, it was felt that such a study might indicate which of two alternative biochemical routes might be responsible for the synthesis of UDP-GlcUA.

Generally, it would appear that, in actively growing tissues where primary cell-wall synthesis predominates, UDP-GlcUA pyrophosphorylase is the dominant enzyme. This is particularly evident in barley and corn seedlings, in which the dehydrogenase is undetectable but the activity of pyrophosphorylase is very high and closely parallels the growth of the young seedling. In such tissues, UDP-GlcUA synthesis presumably proceeds via a pathway from D-glucose 6-phosphate which involves *myo*-inositol as an intermediate. This pathway might operate and be controlled independently of the one leading to UDP-glucose and, hence, polyglucan synthesis. There is no evidence that the pyrophosphorylase is a regulatory enzyme, however. It shows normal Michaelis-Menten kinetics and a magnesium dependency which is related directly to the concentration of substrate UTP and is neither appreciably inhibited nor activated by a range of metabolites that we have tested.

However, when the two enzymes were measured in maple (*Acer rubra*) cambium and sapwood, the specific activity of the dehydrogenase was greater in the woody xylem than it was in the meristematic cambium, while the activity of the pyrophosphorylase fell away markedly during wood development. These observations strongly implicate UDP-glucose dehydrogenase in the synthesis of UDP-GlcUA for polysaccharide biosynthesis during the development of the secondary cell wall of maple.

Acknowledgments

The assistance of Miss Alison Adams and Mr. Bruce Connor in many phases of this study is gratefully appreciated. Fig. 1 is reproduced with the permission of the *Journal of Biological Chemistry*. The work was supported by a grant from the National Science Foundation, GB 23533.

References

1. H. Nikaido and W. Z. Hassid, *Advan. Carbohyd. Chem.* 26, 351, (1971).
2. J. L. Strominger and L. W. Mapson, *Biochem. J.* 66, 567, (1957).
3. D. S. Feingold, E. F. Neufeld, and W. Z. Hassid, *Arch. Biochem. Biophys.* 78, 401, (1958).
4. J. L. Strominger, E. S. Maxwell, J. Axelrod, and H. M. Kalckar, *J. Biol. Chem.* 224, 79, (1957).
5. F. Loewus, S. Kelly, and E. F. Neufeld, *Proc. Natl. Acad. Sci. U.S.* 48, 421, (1962).
6. F. Loewus, *This Symposium, Chap. 1*.
7. R. M. Roberts, *J. Biol. Chem.* 246, 4995, (1971).
8. R. M. Roberts and F. Loewus, *Plant Physiol.* 43, 1710, (1968).
9. K. C. Tovey and R. M. Roberts, *Plant Physiol* 46, 406, (1970).
10. R. M. Roberts, *Exptl. Cell Res.* 46, 495, (1967).

11. F. A. L. Clowes and B. E. Juniper, *Plant Cells*, Blackwell Scientific Publications, Oxford, (1969).
12. V. Ginsburg, *J. Biol. Chem.* *232*, 55, (1958).
13. A. Komogawa and K. Kurahashi, *J. Biochem.* (Tokyo) *57*, 758, (1965).
14. H. Verachtert, P. Rodriguez, S. T. Bass, and R. G. Hansen, *J. Biol. Chem.* *241*, 2007, (1966).
15. J. B. Thornber and D. H. Northcote, *Biochem. J.* *81*, 449, (1961).
16. P. H. Rubery and D. H. Northcote, *Nature 219*, 1230, (1968).
17. J. P. Thornber and D. H. Northcote, *Biochem. J.* *81*, 455, (1961).
18. T. E. Timell, *Advan. Carbohyd. Chem.* *19*, 243, (1964).
19. P. H. Rubery, *Planta 103*, 188, (1972).

BIOSYNTHESIS OF D-XYLOSE AND
L-ARABINOSE IN PLANTS

David Sidney Feingold*
Department of Microbiology, School of Medicine,
University of Pittsburgh
Pittsburgh, Pennsylvania   15213

and

Der Fong Fan
Montefiore Hospital
Pittsburgh, Pennsylvania   15213

Abstract

Uridine diphosphate-D-xylose (UDPXyl) and UDP-
L-arabinose (UDPAra) have been isolated from mung
bean seedlings. UDPXyl is formed *de novo* by action
of UDP-D-glucuronic acid carboxy —lyase in a reaction
which requires NAD and involves UDP-4-keto D-glucuronic acid and UDP-4-keto-D-xylose as intermediates.
UDPAra results from the enzymic 4-epimerization of
UDPXyl. UDPAra pyrophosphorylase from plants catalyzes the formation of UDPAra from UTP and β-L-ara-
binopyranosyl phosphate; synthesis of the latter from
L-arabinose and ATP is catalyzed by a kinase present
in seedlings. UDPXyl functions as donor of the D-
xylopyranosyl moiety to form D-xylans in a reaction

---

*Awardee of the United States Public Health
Service Career Development Grant 1 K3-GM-28,296.

catalyzed by particulate preparations from corn and
other plants; UDPAra probably plays a similar role
in the synthesis of L-arabinose-containing polysaccharides. However, since the latter usually contain
L-arabinofuranosyl moieties, while the L-arabinosyl
moiety of UDPAra is pyranose, a change from the
pyranosyl to furanosyl form must take place prior to
glycosyl transfer. On the basis of the similarities
in the reaction mechanisms of the 4-epimerases and
the carboxyl-lyase it is proposed that these enzymes
have similar structures and that they bear a close
evolutionary relationship to each other.

Symbols

UDPGlc	= uridine 5'-(α-D-gluco-pyranosyl pyrophosphate)
UDPGal	= uridine 5'-(α-D-galacto-pyranosyl pyrophosphate)
UDPXyl	= uridine 5'-(α-D-xylopy-ranosyl pyrophosphate)
UDPAra	= uridine 5'(β-L-arabinosyl pyrophosphate)
UDPGlcUA	= uridine 5'-(α-D-glucopy-ranosyluronic acid pyro-phosphate)
UDPGalUA	= uridine 5'-(α-D-galacto-pyranosyluronic acid pyro-phosphate)
UDP-4-keto-GlcUA	= uridine 5'-(α-D-*xylo*-hexo-pyranosyluronic acid-4-ulose pyrophosphate)
UDP-4-keto-Xyl	= uridine 5'-(α-D-*threo*-pento-pyranosyl-4-ulose pyrophos-phate.

The aldopentoses D-xylose and L-arabinose are
widely distributed in nature. D-xylose is particularly abundant in the polysaccharides of higher plants

(1). It also is found in the polysaccharides of certain yeast-like organisms (i.e., *Cryptococcus laurentii* (2)) and in the lipopolysaccharides of a number of gram-negative bacteria (3). Recently it has become apparent that D-xylose is also present in animal tissues, in which it forms part of the structure of a number of acidic mucopolysaccharide-protein-complexes (4). L-Arabinose, on the other hand, is mainly found in the plant kingdom, where it is an important constituent of cell-wall polysaccharides. Although D-xylose usually occurs as the pyranose form, L-arabinose is mainly found in the furanose structure (1, 5).

The structural similarities of hexosans and pentosans isolated from the same plant source led early workers to postulate that hexosans were converted by oxidation to polyuronides which in turn were decarboxylated to yield pentosans. According to this view, β-1,4-linked D-xylan, which often occurs together with the closely similar hexosan cellulose could be formed from the latter as follows:

$$\text{cellulose} \xrightarrow{\text{(dehydrogenation)}} \text{polyglucuronide} \xrightarrow{\text{(decarboxylation)}} \text{xylan.}$$

An analogous process was assumed to be responsible for the formation of araban from galactan.

It was not until 1942 that Hirst (6) called attention to the inability of the "decarboxylation theory" to explain the formation of araban, in which the L-arabanosyl moieties are furanose, from galactan, in which the D-galactosyl moieties are pyranose. However, he proposed that the conversion of free hexose to pentose by oxidative decarboxylation nevertheless could occur; subsequent polymerization of the monosaccharide residues would then yield the polysaccharide present in such variety in plants.

71

In 1956 Neish (7) and Altermatt and Neish (8) presented evidence to support the "decarboxylation theory" by showing that labeled D-glucose was converted by wheat plants to the D-xylosyl moieties of xylan with loss of C-6 and little randomization of C-1 to C-5. Slater and Beevers (9) further confirmed the decarboxylation hypothesis by demonstrating that corn coleoptile converts D-glucuronolactone to D-xylosyl moieties with loss of C-6. Altermatt and Neish (8) had already suggested that the following stepwise reactions were responsible for the formation of the D-xylosyl residue: D-glucose $\rightarrowtail$ α-D-glucopyranosyl-P → UDPGlc → UDPGlcUA → UDPXyl. The isolation from mung bean seedlings of UDPGlc, UDPXyl (10) and UDPGluUA (11), was consistent with this hypothesis.

In 1958 the conversion of UDPGlcUA to a mixture of UDPGalUA, UDPXyl and UDPAra, catalyzed by a particulate preparation from seedlings of *Phaseolus aureus* was reported (12). Since the enzyme preparation contained both UDPGlcUA-4-epimerase (EC 5.1.3.c) and UDPXyl-4-epimerase (EC 5.1.3.b), it was not clear whether UDPAra arose from the 4-epimerization of UDPXyl or from the decarboxylation of UDPGalUA. Kinetic studies by Feingold et al. (13) with a partially purified enzyme preparation devoid of UDPGluUA-4-epimerase showed that while both UDPXyl and UDPAra were produced from UDPGlcUA, UDPXyl was the first product of UDPGlcUA decarboxylation. Since UDPGalUA was not decarboxylated by the partially purified enzyme, it was concluded that UDPAra was formed from UDPXyl by the UDPXyl-4-epimerase in the enzyme preparation.

UDPGlcUA carboxy-lase (EC 4.1.1.35), the enzyme responsible for the decarboxylation of UDPGlcUA, now has been demonstrated in a wide variety of organisms. In addition to being present in higher plants (13, 14) it has been demonstrated in bacteria (15), green and blue-green algae (16) the yeast-like organism *Crypto-coccus laurentii* (17), hen oviduct (18), cartilage (19), and mast-cell tumors (20). Among higher plants wheat germ was found to be an especially good source of the enzyme (21).

Crude extracts of wheat germ yielded a variety of products from UDPGlcUA, including UDPXyl and UDPAra. However, UDPXyl was the only sugar nucleotide produced by the purified enzyme; in addition the latter did not decarboxylate UDPGalUA (14). These results confirmed previous views that UDPXyl only is produced by the action of UDPGlcUA carboxy-lyase and UDPAra results from the 4-epimerization of UDPXyl.

A number of important aspects of UDPGlcUA carboxy-lyase activity were revealed by comparison of the enzymes of *C. laurentii* and wheat germ. Partially purified *C. laurentii* UDPGlcUA carboxy-lyase was found to have an absolute requirement for NAD; NADH was an inhibitor strictly competitive with NAD. In addition, NAD was very effective in protecting the enzyme from inactivation by heat (17). In contrast to the carboxy-lyase from *Cryptococcus*, the highly purified enzyme from wheat germ was neither activated by NAD nor inhibited by NADH; charcoal treatment or NADase (EC 3.2.2.5) likewise had no effect. However, upon denaturation by heat, the purified enzyme released NAD; suggesting the presence of tightly-bound NAD in the wheat germ enzyme (14). In respect to their involvement with NAD there is a striking similarity between the carboxyl-lyases of wheat germ and *C. laurentii* and the UDPGlc-4-epimerases (EC 5.1.3.2) of *Escherichia coli* (22) and calf liver respectively (23). The *E. coli* enzyme has tightly bound NAD and is not affected by NADH, while the liver enzyme requires NAD and is inhibited by NADH.

It is now known that in the 4-epimerization of UDPGlc (or UDPGal) a hydride ion is transferred from C-4 of the hexosyl moiety of the substrate to the B position of enzyme-bound NAD. This process produces the enzyme-bound intermediate uridine 5'-(α-D-*xylo*-hexpyranosyl-4-ulose pyrophosphate), subsequent reduction of this intermediate by the same hydride ion originally removed from C-4 yields the reaction products, UDPGlc and UDPGal (24).

The similarity in regard to NAD requirements of the 4-epimerases and the carboxy-lyases suggested close similarities in reaction mechanisms also. According to this view, UDP-4-keto-GlcUA, an intermediate analogous to that involved in the action of the 4-epimerase, would be present in the reaction catalyzed by UDPGlcUA carboxy-lyase. Furthermore, mechanistically β-keto acids such as the proposed intermediate are favored for decarboxylation. In such compounds the β-carbonyl group acts as an electron sink to accept the pair of electrons left behind by the leaving carbon dioxide, thus stabilizing the remaining residue. An analogous case is the decarboxylation of acetoacetic acid studied by Westeheimer and co-workers (25). In acetoacetic acid the carbonyl group β to the carboxyl group functions as an electron sink; protonation of the β-carbonyl group enhances its electron withdrawing property and further facilitates the decarboxylation.

More convincing evidence for the involvement of a β-carbonyl carboxylic acid in the action of UDPGlcUA carboxy-lyase was provided by Schutzback and Feingold (26) in studies with the enzymes from *C. laurentii* and wheat germ. UDPGlcUA labeled with ^{3}H at either C-3, C-4, or C-5 of the D-glucuronosyl moiety was converted to UDPXyl with each enzyme. In every case there was complete retention of label; further degradation studies showed that during the decarboxylation there was no migration of ^{3}H from its original position. However, a kinetic isotope effect (V_{3H}/V_{2H}) of approximately 0.35 was observed with the C-4 labeled substrate, but not with the C-3 or C-5 labeled substrates. These results strongly suggest that C-4 is involved in some rate-limiting step of the decarboxylation. A remarkably similar isotope effect has been obtained by other workers in studies with UDPGlc-4-epimerase from E. *coli* (24, 27).

In a separate experiment (see diagram), ^{3}H-label-

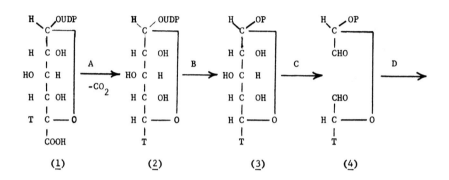

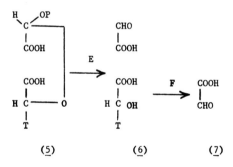

ed UDPXyl (*2*), obtained from the enzymic decarboxylation (A) of UDPGlcUA-5-^{3}H (*1*), was treated with phosphodiesterase (B) to yield ^{3}H-labeled α-D-xylosyl-P (*3*). Periodate oxidation (C) of the latter yielded D-phosphodiglycolic aldehyde, (*4*), from C-1 and C-2 and from C-4 and C-5. The aldehyde was oxidized to D-phosphodiglycolic acid (*5*) with hypobromite (D).

Upon acid hydrolysis (E) of this compound, unlabeled glyoxalic acid (originating from C-1 and
C-2) and labeled glycolic acid (6) (originating from
C-4 and C-5) were released. The latter compound was
converted to unlabeled glyoxalic acid (7) by oxidation with glycolic oxidase from spinach (F). This
enzyme is specific for that hydrogen atom in glycolic acid which is stereospecifically equivalent
to the α-hydrogen of L-lactic acid (28). On the
basis of the stereospecificity of this enzyme, the
absolute configuration at C-2 of $2-^3H$ is "R", which
also must be the absolute configuration at C-5 of the
parent $UDPXyl-5-^3H$. Since the absolute configuration
of $UDPGlcUA-5-^3H$ is "S", complete inversion of configuration must have occurred during the decarboxylation.

Further confirmation of the above finding was
obtained by carrying out the enzymic decarboxylation
with unlabeled UDPGlcUA in ^3H_2O, which procedure
yielded $UDPXyl-5-^3H$. This compound was degraded
exactly as described to give $2-^3H$-glycolic acid. The
specific activity of the latter compound, which was
identical to that of the parent $UDPXyl-5-^3H$, remained
unchanged upon treatment with glycolic oxidase.
These results show that only one proton from the
medium is introduced into the product; they also show
that it is added at C-5 and that the absolute configuration at this C-atom is "R". Parent UDPGlcUA
is "S" at C-5; therefore the proton must be incorporated into the side of C-5 of the glycosyl moiety
opposite the leaving group. Such incorporation requires the presence of an enzyme-bound keto intermediate, since no label was incorporated into UDPXyl
when it was incubated with enzyme in ^3H_2O.

The studies of UDPGlcUA carboxy-lyase in their
entirety are consistent with the following reaction
mechansim. The enzyme would combine with UDPGlcUA

76

$$E-NAD + \underset{K_{-1}}{\overset{K_1}{\rightleftarrows}}$$
$$UDPGlcUA-5T(5S)$$

$$\left[\begin{array}{c} E \\ + \\ NAD \end{array} \quad \text{(I)} \right] \xrightarrow[K_{-2}]{K_2} \left[\begin{array}{c} E \\ NADH \end{array} \quad \text{(II)} \right]$$

$$-CO_2 \downarrow K_3$$

$$\left[\begin{array}{c} E \\ NADH \end{array} \quad \text{(IV)} \right] \xleftarrow[\pm H^+]{K_4} \left[\begin{array}{c} E \\ NADH \end{array} \quad \text{(III)} \right]$$

$$K_{-5} \updownarrow K_5$$

$$\left[\begin{array}{c} E \\ NAD \end{array} \quad \text{(V)} \right] \xrightarrow[K_{-6}]{K_6} \begin{array}{c} E-NAD + \\ UDPXyl-5T(5R) \end{array}$$

in a reversible step represented by K_1 and K_{-1}. The probable rate-limiting step, K_2, would require the participation of NAD in the extraction of a hydride ion from the substrate at C-4. The 4-keto intermediate (II) then would decarboxylate in an irreversible step leaving a carbanion at C-5 (III). During enolization of III, C-5 would assume a planar configuration. Inversion of configuration would occur with the incorporation of the proton at C-5, K_4, to form UDP-4-keto-Xyl (IV). The incorporation of the proton has to occur while the substrate is still enzyme-bound because of the complete stereospecificity of the reaction. If the 4-keto-intermediate (III) were to exist free in the medium, the incorporation of the proton (non-enzyme-mediated) would have been randomized. The reduction at C-4, involving the hydride ion originally removed from C-4, is probably the terminal step in the reaction and appears to be irreversible.

Myo-inositol-2-^{3}H is converted by higher plants to D-xylose-5-^{3}H with inversion of configuration at the labeled-C-atom (29); the following reactions are thought to be involved in this conversion: myo-inositol → D-glucuronic acid →→ α-D-glucopyranosyluronic acid-P → UDPGlcUA → UDPXyl. From the discussion presented above it is clear that the inversion occurs during the decarboxylation step.

UDPAra, the 4-epimer of UDPXyl, is presumably the precursor of glycosidically linked L-arabinose in higher plants, although this has not yet been demonstrated directly. There are two known routes to UDPAra; in one of these free L-arabinose can be converted to UDPAra by the successive action of arabinokinase (30) and UDPpentose phrophosphorylase (31). However, this requires the prior presence of L-arabinose and probably represents a salvage pathway rather than a *de novo* pathway leading from photosynthetically derived D-glucose (32). The *de novo* pathway for synthesis of the L-arabinosyl moiety involves UDPXyl 4-erimerase. The presence of this enzyme was suggested in 1957 when Neufeld et al. (31) showed that crude extracts of *Phaseolus aureus* could interconvert UDPXyl and UDPAra as well as UDPGlc and UDPGal. Subsequently Fan and Feingold (33) partially purified UPDXyl 4-epimerase from wheat germ extracts. The enzyme was shown to have no action on UDPGlc, UDPGlcUA, or dTDPGlc. Furthermore, in contrast to UDPGlc 4-epimerase purified from the same source (34), the UDPXyl 4-epimerase neither requires NAD nor is inhibited by NADH. The equilibrium constant for the reaction UDPAra ⇌ UDPXyl is 1.25. This value contrasts with the value for the analogous reaction UDPGal ⇌ UDPGlc, which is 3.1 (34). The preponderance of UDPGlc in this mixture is thought to reflect the higher energy level associated with the axial hydroxyl at C-4 of UDPGal. It is remarkable that the equilibrium constant should be

so different for the 4-epimerization of the UDP-pen-
toses, which differ from the UDP-hexoses only in
lacking a hydroxymethyl residue. The reason for the
unusually high proportion of UDPAra in the equilibrium mixture is not immediately obvious.

NAD either free or tightly bound has been shown
to be essential for the action of UDPGlc-4-epimerase,
and probably for other nucleoside diphosphate sugars
as well. Constantly accumulating evidence suggests
that all nucleoside diphosphate glycose 4-epimerases
are in fact 4-ulose NAD-oxidoreductases (35, 36).
Thus it is likely that UDPXyl-4-epimerase contains
tightly bound NAD and that UDP-4-keto-Xyl is the
intermediate in the interconversion of UDPXyl and
UDPAra.

The decarboxylation theory as modified by Hirst
has in the main been substantiated for the D-glucose-
-related compounds. One might suspect an analogous
series of reactions to lead from UDPGal to UDPGalUA
and UDPAra. To date evidence for this pathway in
plants has been all but lacking. The one exception
is a report by Katan and Avigad in which presumptive
evidence was presented for the dehydrogenation of
deoxythymidine 5'-(α-D-galactopyranosyl pyrophosphate)
by extracts of sugar beet (37). UPDGalUA carboxy-
lyase, while never demonstrated in plants, has
however, been described in the filamentous procaryo-
tic organism *Ampullariella digitata* (38). In this
organism decarboxylation of UDPGalUA (formed by the
4-epimerization of UDPGlcUA) appears to be the only
biosynthetic route to UDPAra. Although purified
UPDGlcUA carboxy-lyase from wheat germ will not
catalyze the decarboxylation of UDPGalUA, this does
not exclude the presence of UDPGalUA carboxy-lyase
in wheat germ or in other plant tissues, and it still
is a moot point whether UDPAra can be formed in higher plants by the decarboxylation of UDPGalUA.

As was mentioned earlier, glycosidically-bound L-arabinose in plant polysaccharides is predominantly the furanose form, while other glycosyl moieties are mainly pyranose. Particulate preparations from various plants convert UDPXyl to a polysaccharide which is similar to plant xylan and which also contains a small proportion of arabinofuranosyl moieties (39). Since the particulate preparations also contain UDPXyl 4-epimerase, UDPAra as well as UDPXyl must have been present in the reaction mixture. However, the L-arabinosyl moiety of UDPAra formed from UDPXyl by action of UDPXyl-4-epimerase is the pyranose form (40). Therefore, ring contraction attendant upon the formation of L-arabinofuranose moieties of plant polysaccharides either must occur after formation of UDPAra or by some process in which UDPAra is not directly involved, i.e., ring contraction at the polymer level, formation of a nucleoside diphosphate-β-L-arabinofuranose other than the uridine derivative, etc. Direct formation of UDP-β-L-arabinofuranose from UDPAra seems to be the simplest and the most likely possibility. Similar pyranose-furanose ring contractions occur in the formation of UDP-D-apiose from UDPGlcUA in *Lemna minor* L (41), and in the formation of UDP-α-D-galactofuranose in *Penicillium charlesii* (42) (and probably also in *Salmonella typhimurium* (43)). Since the majority (if not all) interconversions of nucleoside diphosphate sugars are pyridine nucleotide-linked oxidoreductions in which a 4-ulose is an obligate intermediate (35, 36), it is possible that UDPAra is converted to UDP-β-L-arabinofuranose by a similar mechansim (possibly involving a 2- or 3-ulose intermediate). Whether this hypothesis has any factual basis must await experimental confirmation.

The similarity of the major types of interconversions of sugar nucleotides, and where known,

of the reaction mechansims involved suggests a close
evolutionary relationship between the several enzymes
which catalyze these reactions (35, 36). If one
assumes that UDPGlc-4-epimerase is the prototype
enzyme of this group, then the other 4-epimerases,
the carboxy-lyases, and related enzymes could be considered to have evolved from and perhaps to be still
related to UDPGlc-4-epimerase. It is significant
that UDPGlc-4-epimerase of *E. coli* catalyzes the
4-epimerization of UDPXyl as well as of UDPGlu (44).
However, in higher plants there exist specific
4-epimerases for UDPGlc, UDPGlcUA, and UDPXyl (13,
33, 34). The bacterial UDPGlc 4-epimerase might
represent a comparatively primitive, relatively un-
specific form of the enzyme, while plant UDPGlc-4-
epimerase could be a more evolved form with greater
specificity for the glycosyl moiety. The plant
enzyme, while utilizing the same reaction mechanism,
and perhaps retaining the structure of the active
site, may differ from the *E. coli* enzyme in its
specificity-related amino acids, making it highly
specific for the D-glucosyl (or D-galactosyl) moiety.
Subsequent evolution of the plant enzyme would lead
to 4-epimerases specific for UDPXyl and UDPGlcUA
(and their 4-epimers).

The intermediate in the decarboxylation of
UDPGlcUA, UDP-4-keto-GlcUA, is identical to the
putative intermediate of the reaction catalyzed by
UDPGlcUA-4-epimerase. The carboxy-lyase therefore
may represent an enzyme which has evolved from the
latter 4-epimerase by acquiring the ability to
facilitate in addition the decarboxylation of the
4-carbonyl intermediate of UDPGlcUA. Similar re-
lationships among the other enzymes of sugar nucleotide interconversions have been recently pointed
out (34, 35).

One cannot but admire the prescience of the
original proponents of the venerable decarboxylation

theory. This hypothesis provided the impetus for work which led to the elucidation of the pathway for pentosan formation and indirectly to our present knowledge of the reaction mechanisms of a number of the participating enzymes. In addition, as discussed in the proceding paragraphs, it also may furnish the key to understanding the evolutionary development of many of the enzymes involved in the various transformations which lead from hexose to pentosan.

Acknowledgment

Research performed in the senior author's laboratory was supported by Grant GM-08820 from the National Institutes of Health, United States Public Health Service and by a grant from the Herman Frasch Foundation.

References

1. P. Albersheim in *Plant Biochemistry*, J. Bonner and J. E. Varner, eds. Academic Press, N.Y., (1965) p. 298.
2. M. J. Abercrombie, J. K. N. Jones, M. V. Lock, M. B. Perry and R. J. Stoodly, *Can J. Biochem. Physiol. 38*, 1617 (1960).
3. W. H. Volk, *J. Bacteriol. 91*, 39 (1966).
4. L. Roden in *Chemical Physiology of Mucopolysaccharides*, G. Quintarelli, ed. Little, Brown and Company, Boston, Mass. (1968) p. 17.
5. G. O. Aspinall, *Polysaccharides*, Pergamon Press, Oxford (1970).
6. E. L. Hirst, *J. Chem. Soc.* p. 70 (1942).
7. A. C. Neish, *Can. J. Biochem. Physiol. 33*, 658 (1955).
8. H. A. Altermatt and A. C. Neish, *Can. J. Biochem. Physiol. 34*, 405 (1956).
9. W. G. Slater and H. Beever, *Plant Physiol. 36*, 146 (1958).

10. V. Ginsburg, P. K. Stumpf and W. Z. Hassid, *J. Biol. Chem. 223*, 977 (1956).
11. J. Solms and W. Z. Hassid, *J. Biol. Chem. 228*, 357 (1957).
12. E. F. Neufeld, D. S. Feingold and W. Z. Hassid, *J. Amer. Chem. Soc. 80*, 4430 (1958).
13. D. S. Feingold, E. F. Neufeld and W. Z. Hassid, *J. Biol. Chem. 235*, 910 (1960).
14. H. Ankel and D. S. Feingold, *Biochemistry 4*, 2468 (1965).
15. D. F. Fan and D. S. Feingold, *Arch. Biochem. Biophys. 148*, 576 (1972).
16. H. Ankel, D. G. Farrell and D. S. Feingold, *Biochim. Biophys. Acta 90*, 397 (1964).
17. H. Ankel and D. S. Feingold, *Biochemistry 5*, 182 (1966).
18. A. Bdolah and D. S. Feingold, *Biochem. Biophys. Res. Commun. 21*, 543 (1966).
19. A. A. Castellani, A. Calatroni and P. G. Righetti, *Ital. J. Biochem. XVI*, 5 (1967).
20. J. E. Silbert and S. DeLuca, *Biochim. Biophys. Acta 141*, 193 (1967).
21. D. S. Feingold, E. F. Neufeld and W. Z. Hassid, *Methods in Enzymology 6*, 782 (1963).
22. D. B. Wilson and D. S. Hogness, *J. Biol. Chem. 239*, 2469 (1964).
23. E. S. Maxwell, *J. Biol. Chem. 229*, 139 (1957).
24. L. Glaser and L. Ward, *Biochim. Biophys. Acta 198*, 613 (1970).
25. F. A. Westheimer and W. A. Jones, *J. Amer. Chem. 63*, 3283 (1941).
26. J. S. Schutzbach and D. S. Feingold, *J. Biol. Chem. 245*, 2476 (1970).
27. G. L. Nelsestuen and S. Kirkwood, *Biochim. Biophys. Acta 220*, 633 (1970).
28. I. A. Rose, *J. Amer. Chem. Soc. 80*, 5835 (1958).
29. F. A. Loewus, *Arch. Biochem. Biophys. 105*, 590 (1964).
30. E. F. Neufeld, D. S. Feingold and W. Z. Hassid, *J. Biol. Chem. 235*, 906 (1960).

31. E. F. Neufeld, V. Ginsburg, E. W. Putman, D. Fanshier and W. Z. Hassid, *Arch. Biochem. Biophys.* **69**, 602 (1957).
32. R. M. Roberts and V. S. Butt, *Planta* **94**, 175 (1970).
33. D. F. Fan and D. S. Feingold, *Plant Physiol.* **46**, 592 (1970).
34. D. F. Fan and D. S. Feingold, *Plant Physiol.* **44**, 599 (1969).
35. O. Gabriel, *Advances in Chemistry*, in press.
36. D. S. Feingold in *Biochemistry of the Glycosidic Linkaage* (R. Piras and H. G. Pontis, eds) Academic Press, N.Y. 1972, p. 79.
37. K. Katan and G. Avigad, *Biochem. Biophys. Res. Commun.* **24**, 18 (1966).
38. D. F. Fan and D. S. Feingold, *Arch. Biochem. Biophys.* **148**, 576 (1972).
39. R. W. Bailey and W. Z. Hassid, *Proc. Natl. Acad. Sci. U.S.* **56**, 1586 (1966).
40. G. O. Aspinall, I. W. Cottrell and N. K. Matheson, *Can. J. Biochem.* **50**, 574 (1972).
41. W. J. Kelleher and H. Grisebach, *Eur. J. Biochem.* **23**, 136 (1971).
42. A. G. Trejo, J. W. Haddock, G. J. F. Chittenden and J. Baddiley, *Biochem. J.* **122**, 49 (1971).
43. M. Sarvas and H. Nikaido, *J. Bacteriol.* **105**, 1063 (1971).
44. H. Ankel and U. S. Maitra, *Biochem. Biophys. Res. Commun.* **32**, 1526 (1968).

OCCURRENCE AND METABOLISM OF D-APIOSE
IN *LEMNA MINOR**

Paul K. Kindel

Department of Biochemistry,
Michigan State University,
East Lansing, Michigan   48823

Abstract

Greater than 80% of the D-apiose of *Lemna minor*
occurs in the cell wall. Of the total D-apiose present in the cell wall, a minimum of 20% occurs in a
single type of cell wall polysaccharide. This type
of polysaccharide, termed apiogalacturonan, accounts
for at least 14% of the dry weight of the cell wall
of the plant. Apiogalacturonan from *L. minor* is not
completely homogeneous but can be separated into
sub-groups. The separated apiogalacturonans all have
the same basic structure. They consist of a galacturonan backbone to which are attached apiobiose side
chains. Apiogalacturonans are degraded by unusually
mild conditions to galacturonan and apiobiose. At
pH 4.5 and 100° degradation is virtually complete
after 3 hr.

A method for synthesizing uridine 5'-(α-D-[U-^{14}C]
apio-D-furanosyl pyrophosphate) from uridine 5'-(α-D-
[U-^{14}C]glucopyranosyluronic acid pyrophosphate) in 60

---

*This research was supported by Research Grants
AM-08608 and AM-14189 from the National Institutes of
Health and by the Michigan Agricultural Experiment
Station.

to 63% yield was developed. The stability of uridine 5'-(α-D-[U-^{14}C]apio-D-furanosyl pyrophosphate) at pH 8.0 and 80°, at pH 8.0 and 25°, at pH 8.0 and 4°, and at pH 3.0 and 40° was determined. Uridine 5'-(α -D-[U-^{14}C]apio-D-furanosyl pyrophosphate) was shown to be involved in the biosynthesis of the apiobiose component of either the apiogalacturonans or of a compound posessing some of the same properties as the apiogalacturonans.

Symbols

UDPApi   = Uridine 5'-(α-D-apio-D-furanosyl pyrophosphate)

UDPXyl   = Uridine 5'-(α-D-xylopyranosyl pyrophosphate)

UDPGluUA   = Uridine 5'-(α-D-glucopyranosyluronic acid pyrophosphate)

UMP   = Uridine 5'-phosphate

UDP   = Uridine 5'-diphosphate

DEAE   = Diethylaminoethyl

The branched-chain sugar D-apiose (3-C-hydroxy-methyl-*aldehydo*-**D**-*glycero*-tetrose) is a natural constituent of a large number of different plants (see references in 1) where it has been found in two types of compounds, glycosides and polysac-charides. The isolation and characterization of cell wall polysaccharides of *Lemna minor* containing D-apiose, the synthesis and properties of UDPApi, and the involvement of this sugar nucleotide in apiogalacturonan biosynthesis are discussed in this paper. *L. minor* was used in these studies because of its high D-apiose content which is approximately 6% on a dry weight basis.

Isolation and Characterization of Cell
Wall Polysaccharides of *L. minor*
Containing D-Apiose

Cell walls were isolated from *L. minor* and were found to contain, at a minimum, 83% of the total D-apiose present in the whole plant. A procedure was developed for extracting polysaccharides containing D-apiose from these cell walls. The procedure consisted basically of two steps, an extraction with water followed by an extraction with ammonium oxalate. The material solubilized by ammonium oxalate constituted 14% of the cell wall and contained 20% of the D-apiose of the cell wall. These latter two values are minimal values since this material was taken through a one-step purification procedure after being extracted and before being weighed. The material solubilized by ammonium oxalate was termed apiogalacturonan since it was found to be a single type of polysaccharide that contained D-apiose and D-galacturonic acid. Apiogalacturonan could be separated into sub-groups by NaCl fractionation and by DEAE-Sephadex column chromatography, indicating that is was not homogeneous. As used here homogeneous means that the individual polysaccharide molecules have chemically identical repeating units but are not necessarily the same molecular weight. Subsequent analysis of the subgroups revealed that their D-apiose content ranged from 7.9 to 38.1%. These and other data established the existence of a family of apiogalacturonans. The variable D-apiose content is responsbile for separation of apiogalacturonan into sub-groups.

The insoluble material remaining after the two-step extraction procedure mentioned in the preceeding paragraph was termed the cell wall residue. Sixty-nine per cent (69%) of the starting cell wall material and 76% of the total D-apiose present in the cell wall was recovered in this residue. The identity of the compounds containing D-apiose still

87

present in the cell wall residue is at present unknown.

Partial characterization has revealed that in spite of their variable D-apiose content all apiogalacturonans possess the same basic structure; namely, a galacturonan backbone to which are attached side chains consisting of two D-apiose residues. Side chains of one D-apiose residue may be present in small amounts. Apiogalacturonans undergo a novel degradation reaction. When heated at pH 4.5 and 100° for 3 hr they release their side chains intact resulting in the production of a new disaccharide and galacturonan. The new disaccharide was named apiobiose.* More complete descriptions of the above work are given in Hart and Kindel (1, 2). Apiogalacturonans have also been isolated from L. *minor* by Beck (3).

Synthesis and Properties of UDPApi

The biosynthesis of D-apiose from UDPGluUA by cell-free extracts of L. *minor* has been described (4-6). The reaction involved is the conversion of UDPGluUA to UDPApi and presumably CO_2. The enzyme catalyzing this reaction has been purified 55-fold from L. *minor* (7)** and was given the common name UDPGluUA cyclase (8). The purified enzyme still contained UDPGluUA carboxy-lyase activity. In order to determine if UDPApi was involved in apiogalacturonan biosynthesis a method for synthesizing this sugar nucleotide was developed. The method is based on the use of purified UDPGluUA cyclase. With this method

---

*In a previous paper (2) the disaccharide was named apibiose. The name apiobiose is more in accord with current nomenclature practices and will be used henceforth for this disaccharide.

**D. L. Gustine and P. K. Kindel, in preparation.

UDP[U-^{14}C]Api plus UDP[U-^{14}C]Xyl was synthesized in 85 to 90% yield. Since the ratio of UDP[U-^{14}C]Api to UDP[U-^{14}C]Xyl was normally about 7:3, the yields of UDP[U-^{14}C]Api ranged between 60 and 63%.

To ensure that the conditions for synthesizing and storing UDP[U-^{14}C]Api were optimal, the stability properties of this sugar necleotide were determined. The stability properties of UDP[U-^{14}C]Api are of interest in themselves since sugar nucleotides with the sugar portion in the furanose ring form are novel. The only other sugar nucleotide reported with this type of structure is UDP-D-galactofuranose (9) and its stability properties have not yet been described. Particularly noteworthy for UDP[U-^{14}C]Api is its extreme lability in alkali. In alkali UDP[U-^{14}C]Api is degraded to UMP and α-D-[U-^{14}C]apio-D-furanosyl cyclic-1:2-P. At pH 8.0 and 80°greater than 90% of the UDP[U-^{14}C]Api was degraded to these two compounds in less than 2 min. The half-life periods of UDP[U-^{14}C]Api at pH 8.0 and 80°, at pH 8.0 and 25°, and at pH 8.0 and 4° were 31.6 sec, 97.2 min, and 16.5 hr, respectively. As was expected UDP[U-^{14}C]Api was readily hydrolyzed by acid. After 16 min at pH 3.0 and 40°, greater that 90% of the UDP[U-^{14}C]Api was hydrolyzed to UDP and [U-^{14}C]apiose. The half-life period of UDP[U-^{14}C]Api under these conditions was 4.67 min. A more extensive description of the method for synthesizing UDP[U-^{14}C]Api as well as additional information on the properties of UDP[U-^{14}C]Api are given in Kindel and Watson (10).

Biosynthesis of Apiogalacturonans of L. *minor*

Information on the general aspects of plant cell wall polysaccharide biosynthesis as well as specific details of the biosynthesis of a particular plant cell wall heteropolysaccharide should result from a study of apiogalacturonan biosynthesis in L. *minor*. There are several advantages in studying the bio-synthesis of this particular polysaccharide to obtain such information. First, apiogalacturonans have been demonstrated to be present in the cell wall of *L. minor*. Second, the basic structure of the apiogalacturonans is known. Third, the enzymes involved in the biosynthesis of the apiogalacturonans should be relatively easy to detect since these polysaccharides constitute a substantial portion of the cell wall of *L. minor*. Finally, procedures for isolating, characterizing, and degrading the apio-galacturonans have already been developed.

It is reasonable to expect that sugar nucleo-tides, either directly or indirectly, are the precur-sors of the two types of sugar residues present in the apiogalacturonans. With UDP[U-^{14}C]Api now avail-able experiments were initiated to determine if this sugar nucleotide was involved in apiogalacturonan biosynthesis. A particulate enzyme preparation was prepared from *L. minor*. *L. minor*, grown as described previously (10), was harvested by filtration through cheesecloth, washed with distilled water, blotted dry, and weighed. All subsequent steps were performed at 4°. The plants (10 g) were suspended in 10 ml of 0.05 M sodium phosphate buffer, pH 7.3, containing 1 mM $MgCl_2$ and 1% (w/v) bovine serum albumin and ground with a mortar and pestle. Two additional volumes of the above buffer were added to the thick suspension and the resulting slurry was further homogenized with a motor-driven homogenizer. The homogenate was filtered through 4 layers of cheese-cloth. The filtrate was centrifuged at 1000g for 10 min. The supernatant solution was centrifuged at 35,000g for 30 min and the precipitate was gently resuspended with a glass homogenizer in 0.05 M sodium phosphate buffer, pH 6.8, containing 0.4 M sucrose and 1% bovine serum albumin. The precipitate from the 35,000g centrifugation was used as the source of the enzyme(s) in the experiments described. In this paper the suspended precipitate is termed the particulate enzyme preparation. An incubation mixture was prepared which contained the particulate

enzyme preparation (0.05 ml), UDP[U-^{14}C]Api
(16,630 dpm, 37.8 pmoles), UDP[U-^{14}C]Xyl (6695 dpm,
15.2 pmoles) and UDP[U-^{14}C]GluUA (192 dpm, 0.36 pmole)
in a final volume of 0.08 ml. The mixture was
incubated 30 min at 25° and then 3.0 ml of 75%
methanol containing 1% KCl was added and the suspension was centrifuged. The supernatant solution was
discarded and the precipitate was extracted in order
with aqueous 75% (v/v) methanol containing 1% (w/v)
KCl, methanol, aqueous 75% methanol containing 1% KCl
and water. Except for the water extractions each
extraction was for 5 min at 25°. The water extractions were for 15 min at 25°. Each type of extraction was continued until no further radioactivity was
extracted or until the amount of radioactivity extracted each time had reached a low, constant value.
The residue remaining was suspended in aqueous 2% (w/v)
NaOH and the suspension was spotted on Whatman No. 3
MM paper. The paper containing the residue was placed
in a solution of 2,5-bis[2-(5-*tert*-butylbenzoxazolyl)]
thiophene in reagent grade toluene (4 g/l.) and assayed for radioactivity with a Packard Tri-Carb liquid
scintillation counter, Model 3310. The incorporation
of radioactivity into the residue in 7 experiments
ranged from 17 to 25%. The data were highly repro-
ducible. In two of these experiments a second incubation mixture was prepared identical to the first
except that the particulate enzyme preparation was
heated 3 min at 100° before being added to the
UDP[U-^{14}C]Api. In these two experiments the incorporation of radioactivity into the residue was 0.02
and 0.05%. The above type of experiment was carried
out on a larger scale. The amount of particulate
enzyme preparation and UDP[U-^{14}C]Api used was 10 and
25 times greater, respectively, than that used in the
above experiments. In the large-scale experiments the
residue obtained following the above methanol and
water extractions was further extracted with aqueous
0.5% (w/v) ammonium oxalate for 25 min at 25° (2 X)
and then with water for 25 min at 25° (2 X). The
material solubilized by these extractions was combined

and found to contain 5.9% of the starting radioactivity. The material extracted by the ammonium oxalate
treatment in 14.8 ml was dialyzed against 1 1. quantities of water for a total time of 3 hr with 3
changes of the water. The non-dialyzable material
contained 5.0% of the starting radioactivity. The
values for the per cent incorporation of radioactivity
were calculated on the basis of the starting radioactivity in UDP[U-^{14}C]Api alone since only D-[U-^{14}C]
apiose could be detected after acid hydrolysis of the
initial residue obtained in the above extraction procedure. It is possible that D-[U-^{14}C]xylose was
incorporated into this residue but in quantities too
small to be detected.

The non-dialyzable material was adjusted to pH
4.0 with 0.02 M potassium phosphate buffer, pH 4.0,
and heated at 100° for 3 hr. The hydrolysate was
spotted on Whatman No. 3 MM paper prewashed with
0.1 M citric acid and water. The paper was developed
with ethyl acetate:water:acetic acid:formic acid
(18:4:3:1, by vol.) for 6.5 hr and then scanned for
radioactivity with a Packard radiochromatogram
scanner, Model 7201. Radioactive material with R_F
of authentic apiobiose was obtained (Fig. 1). When
the non-dialyzable material was chromatographed on
paper without being heated for 3 hr at pH 4.0 and

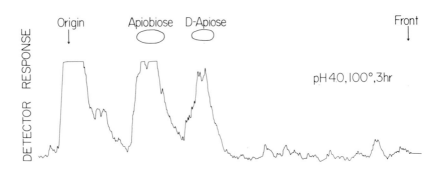

Fig. 1 Radiochromatogram scan of heated non-
-dialyzable material.

100°, radioactive material could only be detected at
the origin. Under the above chromatography conditions
authentic D-xylose migrated faster than apiobiose
but still overlapped the disaccharide. Therefore it
was necessary to establish, that radioactive apiobiose
and not D-$[U-^{14}C]$xylose was present on the chormato-
gram. The radioactive material with a R_F of apiobiose was hydrolyzed with 0.05 M H_2SO_4 for 1 hr at
100°. The neutralized hydrolysate was chromatographed
in the above solvent for 12 hr. The only ratioactive
material observed on the chromatogram had a R_F of
D-apiose and was widely separated from authentic
D-xylose and apiobiose. This showed that the
radioactive material with a R_F of apiobiose was completely converted to D-$[U-^{14}C]$apiose. In addition,
the radioactive material migrated on paper with the
same R_F as authentic apiobiose in two other solvents.
Furthermore, the conditions of hydrolysis (pH 4.0, 100°
3 hr) should not be vigorous enough to release free
D-xylose from glycosidic linkages unless these linkages were unusually acid labile. These results showed
that the radioactive material with a R_F of apiobiose
was indeed radioactive apiobiose and that D-$[U-^{14}C]$
xylose was not present on the original chromatogram.

The above data showed that the D-$[U-^{14}C]$apiose
moiety of UDP$[U-^{14}C]$Api was used to synthesize the
apiobiose component of either the apiogalacturonans
or of a compound having some of the same properties as
those possessed by the apiogalacturonans. This biosynthetic process is currently being examined in
greater detail.

The cell-free system described in this report
appears to be an excellent system for securing information on the biosynthesis of one of the major types
of plant cell wall polysaccharides, the pectic substances. A number of glycosyl transfer reactions that
are considered to be involved in cell wall biogenesis
have been previously observed in cell-free systems
from plants. However, the physiological significance

of these reactions is uncertain since, in almost every
case, it was not determined if the reaction product
was a normal constituent of the cell wall of the
plant under investigation. This uncertainty is
eliminated in studying apiogalacturonan biosynthesis
in cell-free extracts from L. *minor* since apiogalacturonan has been shown to be a normal, and in
fact substantial, constituent of the cell wall of
this plant.

References

1. D. A. Hart and P. K. Kindel, *Biochem. J. 116*,
   569 (1970).
2. D. A. Hart and P. K. Kindel, *Biochemistry 9*, 2190
   (1970).
3. E. Beck, Z. *Pflanzenphysiol. 57*, 444 (1967).
4. H. Sandermann, Jr., G. T. Tisue and H. Grisebach,
   Biochim. Biophys. Acta 165, 550 (1968).
5. D. L. Gustine and P. K. Kindel, J. *Biol. Chem.
   244*, 1382 (1969).
6. J. Mendicino and R. Hanna, J. *Biol. Chem. 245*,
   6113 (1970).
7. D. L. Gustine, Ph.D. Thesis, Michigan State
   University, East Lansing, Michigan, 1969.
8. P. K. Kindel, D. L. Gustine, and R. R. Watson,
   Fed. Proc. Fed. Amer. Soc. Exp. Biol. 30, 1117
   (1971).
9. A. G. Trejo, G. J. F. Chittenden, J. G. Buchanan,
   and J. Baddiley, *Biochem. J. 117*, 637 (1970):
   A. G. Trejo, J. W. Haddock, G. J. F. Chittenden,
   and J. Baddiley, *Biochem. J. 122*, 49 (1971).
10. P. K. Kindel and R. R. Watson, *Biochem. J.*
    (in press).

CARBOHYDRATE POLYMERS OF PLANT CELL WALLS

Gerald O. Aspinall

Department of Chemistry
York University
Downsview, Ontario

Abstract

The chemistry of plant cell wall polysaccharides is reviewed with particular reference to the hemicelluloses, and to pectins and associated neutral polysaccharides. Special attention is given to the structural role of neutral sugar units in pectins and to the problem of polysaccharide homogeity.

The carbohydrate polymers of plant cells fall into a number of groups which are summarized in Fig. 1. Brief comment will be made on each of these groups but particular attention will be given to the chemistry of the pectic substances and associated polysaccharides.

The covalent structure of cellulose was established some forty years ago (1). Since that time fundamental investigations have been directed to obtaining accurate values for the molecular size of undegraded native cellulose and to the determination of intramolecular and inter-chain hydrogen bonding and other conformational factors which influence secondary and tertiary structure (2). The biosynthesis of cellulose in primary and secondary walls

<u>Cellulose</u>

<u>Hemicelluloses</u>	Xylans
	Glucommannans

<u>Pectic Substances</u>	Galacturonans
	Arabinans
	Galactans and/or Arabinogalactans I* (or II)

<u>Other Polysaccharides</u>	Arabinogalactans II*
	Fuco-(or galacto-)xyloglucans

<u>Glycoproteins</u>

*Arabinogalactans of type I are characterized by essentially linear chains of (1→4) linked chains of β-$\underline{D}$-galactopyranose residues, whereas those of type II contain highly branched interior chains with (1→3) and (1→6) inter-galactose linkages.

Fig. 1 Carbohydrate polymers of plant cell walls.

can occur simultaneously, but differences in mechanisms are implied since the molecular size of the latter is constant [$\bar{P}_W$ 13,000 to 16,500 according to the source] and independent of the extent of conversion, whereas that of the former is of lower magnitude and of non-uniform distribution. It has been suggested that the molecular size of secondary wall cellulose must be genetically determined (3), but that morphology and crystal structure are dependent of the biosynthetic process, and are dependent only on the molecular properties of cellulose itself.

From extensive investigations during the 1950s and early 1960s it became apparent that most of the

hemicelluloses of lignified tissues fall into two
broad groups, namely, the xylans and the glucomannans
(4). Within each group the various polysaccharides
contain interior chains of sugar residues in which
only small differences in detailed structure have
been detected. In contrast, quite considerable
variations in structure have been observed in respect
of the number, nature, and mode of attachment of
sugar residues and other substituents, e.g. O-acetyl
groups, which are attached as side-chains. These
structural variations between polysaccharides of the
same general family result in differences in physical
properties and presumable reflect differences in biological function. Generalized structures for polysaccharides of these two groups are shown in Figs.
2 and 3.

Polysaccharides of the xylan group, invariably,
and of the glucomannan group, sometimes, are components of secondary cell walls. The pectic substances,
in contrast, are mainly components of primary walls.
However, before proceeding to a detailed consideration of these latter polysaccharides, reference will
be made to two other types of polysaccharide which
interestingly are elaborated together with pectin
by suspension cultured sycamore cells (5). Arabinogalactans of type II are most commonly found in
coniferous woods (4) and are particularly abundant in
larches where the polysaccharide is present in the
lumen of tracheids and ray cells rather than as a
cell wall component (6). Arabinogalactans of this
type contain highly branched structures in which
β-D-galactopyranose residues are mutually joined by
($1\rightarrow 3$) and ($1\rightarrow 6$) linkages, the former predominantly
in interior and the latter mainly in exterior
chains. L-Arabinofuranose, and to a smaller extent
L-arabinopyranosyl residues, terminate some of the
outer chains. A typical structure for an arabinogalactan of type II is shown in Fig. 4. Some
arabinogalactans of this type also contain terminal
units of D-glucuronic acid (or its 4-methyl ether).

97

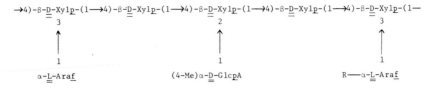

Fig. 2 General formula for xylans, where (a) not all types of side-chains are necessarily present in a particular polysaccharide (b) R represents one or two further sugar residues in a few polysaccharides, and (c) in xylans from hard woods O-acetyl groups are attached to 0-2 or 0-3 of some β-D-xylopyranose residues.

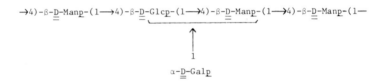

Fig. 3 General formula for glucomannans, where (a) some polysaccharides carry some α-D-galactopyranose residues as single unit side-chains joined usually by (1→ 6) linkages and most frequently to D-mannopyranose residues, and (b) in glucomannans from soft woods O-acetyl groups are attached to D-glucose or D-mannose residues.

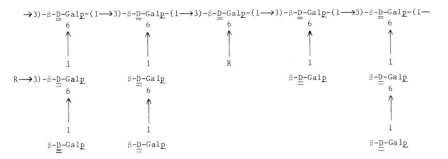

Fig. 4 Representative structure for arabino-galactans of type II. The distribution of $(1 \rightarrow 3)$ and $(1 \rightarrow 6)$ linkages is largely as indicated but not necessarily in a completely non-ramified comb-like structure. The sites of attachment of side-chains [R = L-Araf-(1- or β-L-Arap-(1 → 3)-L-Araf(1-] have not yet been determined with certainty.

There is no clear line of structural demarcation between the coniferous wood arabinogalactans and some of the exudate gums such as gum arabic and other *Acacia* gums (7), and mesquite gums (8).

The xyloglucans contain a cellulose-like β-D-glucan core to which a single unit α-D-xylopyranose residues are attached as side-chains with some of these units being further substituted by L-fucopy-ranose or D-galactopyranose residues (9). They have been found in the seeds of many plants and their pos-sible role as cell wall components is unknown, but the isolation of a fucoxyloglucan from suspension cultured sycamore cells (5) (see Fig. 5 for a par-tial structure) suggests that they may have a spe-cial significance. The structures of the xyloglucans imply that they may be further transformation pro-ducts of cellulose.

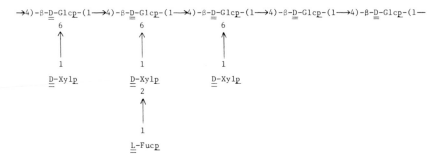

Fig. 5 Partial structure of fucoxyloglucan from suspension cultured sycamore cells

Recent investigations on pectins and certain exudate gums have shown that galacturonans and galacturonorhamnans, like the xylans and glucomannans, may be classified in terms of the basal chains of sugar residues which form the interior chains of the polysaccharide structures, but to which a variety of other sugar residues may be attached in side-chains. Two questions, however, neither of them unique to pectin chemistry, present themselves in a particularly acute form, namely, 'How does one know that a pectin preparation is sufficiently chemically homogeneous for meaningful structural studies to be undertaken?' and 'If a polysaccharide preparation is heterogeneous, are all the components of genuine natural occurrence or do some result from inadvertent degradation during isolation?'. These questions may be considered by looking at four roughly chronological stages in the development of the structural chemistry of pectic substances (Fig. 6).

1. The homopolysaccharide era

2. The recognition of pectic acids as
   heteropolysaccharides - the structural
   role of neutral sugars

3. The degradation of pectins - occurrence
   during isolation and fractionation

4. The problem of polysaccharide
   heterogeneity

Fig. 6 Stages in the development of the structural chemistry of pectic substances

It is of interest to recall in this Symposium devoted to the biogenesis of plant cell wall polysaccharides, that one of the driving forces for the first detailed structural investigation of polysaccharides of this group arose from an interest in the phylogenetic relationship between the hexose, D-galactose, the hexuronic acid D-galacturonic acid, and the pentose L-arabinose. Were these stereochemically-related sugars converted one into the other by sequential oxidation and decarboxylation? In the event structural studies on homopolysaccharides based on these three respective sugars clearly showed that polymers of one sugar could not be readily converted into polymers of another since the polymers differ from each other in one or more of the following respects, configuration of glycosidic linkage, type of linkage or ring size. It is, of course, now well known that sugar interconversions take place mainly at the sugar nucleotide level, prior to polymerization. It is noteworthy, however, that no satisfactory explanation has yet been forthcoming for the origin of L-arabinofuranose residues in Nature. Enzymes mediating the 4-epimerization of UDP-D-xylopyranose furnish UDP-L-arabinopyranose (10).

Hirst and Jones, in about 1940, by the selection of suitable pectins and where necessary employing conditions leading to the selective degradation of an unwanted component, were able to isolate poly- saccharide preparations in which each of the three main pectic sugars were individually present, and whose structure are shown in Fig. 7. This important development merely established the presence of poly- meric units of the three sugars in pectic complexes and was not regarded by the authors as proof that pectic substances were simple mixtures of the three homopolysaccharides (11). Nevertheless the con- cept of the so-called pectic triad of galacturonan, galactan, and arabinan has become firmly established as part of the dogma of polysaccharide mythology.

→4)-β-D-Galp-(1→4)-β-D-Galp-(1→4)-β-D-Galp-(1→4)-β-D-Galp-(1—

→4)-α-D-GalpA-(1→4)-α-D-GalpA-(1→4)-α-D-GalpA-(1→4)-α-D-GalpA-(1—

```
→5)-α-L-Araf-(1→5)-α-L-Araf-(1→5)-α-L-Araf-(1→5)-α-L-Araf-(1—
           3                            3
           ↑                            ↑
           |                            |
           1                            1
       α-L-Araf                     α-L-Araf
```

Fig. 7   Galactan, galacturonan (pectic acid) and arabinan.

The next stage in the development of pectin chemistry came in the late 1950s and early 1960s when polysaccharide preparations particularly those rich in galacturonic acid, were isolated under much milder conditions, and despite extensive fractionation pro- cedures, failed to yield pure galacturonans. It is now apparent that the majority of polysaccharides

rich in galacturonic acid contain significant
amounts of neutral sugar substituents and that pure
galacturonans are of infrequent occurrence. Some
typical pectin compositions are shown in Fig. 8. The
best authenticated example of such pure galacturonans
is that from sunflower heads (12). Another galacturonan has been isolated as a sub-fraction from the
pectin of Amabilis fir bark (13), but for reasons
given later the conditions of isolation leave doubt
as to its natural occurrence. Pectins, e.g. from
sugar-beet, may contain O-acetyl groups which may
markedly modify their gelling properties.

Source of Pectin	Sugar Constituents						
	GalA	Gal	Ara	Rha	Xyl	Fuc	Apiose
Apple	87%	1%	9%	1%	1%		
Lemon-peel	84%	+	+	+	+	tr	
Alfalfa	81%	+	+	+	+	tr	
Amabilis fir bark	77%	6%	17%	+			
Duckweed (Lemma minor)	53-80%						38-7%
Mustard cotyledons	28%	8%	44%	3%	17%		
Soybean cotyledons	24%	+	+	+	+	+	
Tragacanthic acid	43%	4%	tr	tr	40%	10%	

Fig. 8 The composition of some pectins and
tragacanthic acid.

We may turn now to the structural role of neutral sugars in pectins. The role of L-rhamnose may
be assessed separately, since as far as known in
pectins, this sugar occurs only in the interior
chains, whereas residues of all other neutral sugars
are encountered in the exterior chains only. Some
of the ways in which galacturonan chains in pectins
and certain exudate gums may be interrupted with
varying degrees of frequency and regularity by residues of L-rhamnose are illustrated in Fig. 9.
Model building computations for galacturonans indicate that the insertion of L-rhamnose units inter-

→4)-α-$\underline{\underline{D}}$-Gal$\underline{p}$A-(1→4)-α-$\underline{\underline{D}}$-Gal$\underline{p}$A-(1→4)-α-$\underline{\underline{D}}$-Gal$\underline{p}$A-(1—

Galacturonan

-[→4)-α-$\underline{\underline{D}}$-Gal$\underline{p}$A-(1-]$_x$→2)-$\underline{L}$-Rha$\underline{p}$-(1—[→4)-α-$\underline{\underline{D}}$-Gal$\underline{p}$A-(1-]$_y$-

Tragacanthic acid

-[→4)-GalA-(1-]$_x$→4)-GalA-(1→2)-Rha-(1→4)-GalA-(1→2)-Rha-(1→2)-Rha-(1—[→4)-GalA-(1-]$_y$-

Pectins from lemon-peel, alfalfa, and soybeans*

→4)-GalA-(1→2)-Rha-(1→4)-GalA-(1→2)-Rha-(1→4)-Gal-(1→4)-GalA-(1→4)-Gal-(1→4)-GalA-(1-

Karaya gum (from Sterculia urens)*

*In the absence of enantiomeric and configurational prefixes it is assumed (and in most cases proven) that residues are those of α-D-galacturonic acid, D-galactose, and L-rhamnose. All residues are in the pyranose form.

Fig. 9 Interior chains in pectins and structurally-related gums.

---

rupts the tendency to form ordered chain conformations (14). These results suggest a strong analogy with other biological-network polysaccharides where the interspersion of anomalous units causes 'kinking' of otherwise regular chains.

Some of the most important evidence concerning the nature and modes of attachment of neutral sugar residues in side-chains to galacturonans or galacturonorhamnans, came not from pectins themselves, but from tragacanthic acid, the major acidic polysaccharide component of gum tragacanth (15). The side-chains in this polysaccharide of D-xylopyranose residues alone or with appended D-galactopyranose or L-fucopyranose units were then regarded as uncharacteristic of pectins, but have subsequently been found as important structural units in pectins from tissues with potential for rapid enlarge-

104

ment and/or rapid differentiation, such as pollen
(16) and cotyledons from soybeans (17) and white mustard seeds (18). These units are also as minor
structural features in more typical pectins such as
those from lemon peel (19) and alfalfa (20). Side-
chains containing D-galactopyranose and L-arabinofuranose residues are more characteristic of typical
pectins. Information on their mode of attachment is
at present more limited. Fig. 10 summarizes the
types of side-chains for which there is good evidence.

L-Araf-(1-	Alfalfa			
[L-Araf]$_n$—*		Mustard seeds		
[β-D-Galp]$_n$—*			Soybeans	
β-D-Xylp-(1-	Alfalfa	Mustard seeds	Soybeans	Lemon-peel
β-D-Galp-(1→2)-D-Xylp-(1-			Soybeans	
α-L-Fucp-(1→2)-D-Xylp-(1-			Soybeans	
D-Apif-(1→3)-D-Apif-(1-				Duckweed

* The multiple sugar residues in these pectin
sub-units are mutually linked in the same manner as
in arabinans and (arabino)galactans I respectively.

Fig. 10 Side-chains in pectins

In the light of the demonstration that the majority of pectins and pectic acids contain neutral
sugars as integral constituents it becomes necessary
to re-examine the homopolysaccharide concept of the
pectic traid, and to consider how far the isolation
of individual homopolysaccharides may be a con-

sequence of methods used in the isolation and fractionation of the polysaccharides themselves. It is
of course obvious that strongly acidic conditions
must be avoided since these may lead to the hydrolysis of the more acid-labile glycosidic linkages, especially those of L-arabinofuranose and even more of
D-apiofuranose residues. In certain structural
situations the cleavage of only a single glycosidic
bond could result in the separation of an arabinan
sub-unit from the parent complex heteropolysaccharide.

Pectins, particularly those with high proportions of esterified D-galacturonic acid residues,
are also susceptible to base-catalysed degradation
leading to the cleavage of glycosidic linkages with
the formation of unsaturated hexuronic acid units and
to the formation of new reducing groups. Under such
alkaline conditions the β-elimination reaction com-
petes with ester hydrolysis, which latter reaction
furnishes the much less base-sensitive pectic acid.
The elimination, however, may also take place under
ostensibly neutral conditions, e.g. by heating in
phosphate buffer at pH 6.8 (21), and indeed in the
presence of appropriate ionic species may occur slowly
at a pH as low as 4.2 (22). In a polysaccharide of
regularly repeating structure, this type of degradation would merely result in depolymerization with the
formation of chemically similar fragments. Barrett
and Northcote (23), however, have shown that apple
pectin, when heated in phosphate buffer at pH 6.8,
is degraded with the formation of two quite distinct
polysaccharide fractions (Fig. 11). This degradation
thus provides evidence that neutral sugar residues
are distributed irregularly or at least in widely
spaced blocks along the macromolecular chain.

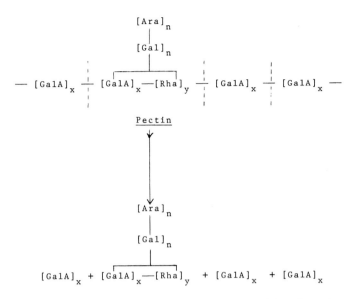

Fig. 11 Formation of two chemically distinct degraded polysaccharides on random degradation of pectin containing an uneven distribution of neutral sugars in the structure.

Bearing in mind these types of degradation, procedures have been developed for the isolation and fractionation of pectic substances which minimize the possibility of alterations in the structures of the native biopolymers. Thus, we have obtained from soybean cotyledons three polysaccharides, an arabinan, an arabinogalactan of type I, and an acidic polysaccharide, or pectic acid, whose method if isolation and whose structures together preclude the possibility of any one polysaccharide arising from another by inadvertant degradation (24). Similarly, graded extraction and fractionation of polysaccharides from lemon peel have furnished two pectins (C and L), an arabinan, an arabinogalactan of type I, and an acidic polysaccharide of unproven homogeneity containing structural features of arabinogalactan

type II (19, 22). The isolation procedure is summarized in Fig. 12. From these results and from
those of several other investigations in our own
and other laboratories we may turn our attention
firstly to the neutral (or only slightly acidic)
polysaccharides and then to the highly acidic
polysaccharides.

The isolation from the same source of both
arabinan and arabinogalactan shows quite clearly
that L-arabinose may be a constituent of both homo-
and heteropolysaccharides. The presence of only
limited branching through arabinose residues
in the arabinogalactans shows that the highly branched arabinans must be distinct species which cannot
arise from partial degradation of the more complex
neutral polysaccharide. The galactans or arabinogalactans which are found in association with pectins are most commonly those based on linear chains
of (1→ 4)-linked β-D-galactopyranose residues. It
is now apparent, however, that composition alone does
not define structure since arabinogalactans of type
II, similar in composition but of quite different
structure to those of type I, have been found in
association with pectins.

In order to assess fully the structures of
pectins, account must be taken of differences in the
degrees of esterification of D-galacturonic acid
residues in addition to variations in sugar composition. In certain cases, e.g. sugar-beet pectin, it
is also necessary to recognise the presence of O-
acetyl subsuituents. Lemon-peel provided a convenient source from which to isolate a pectin under
the mildest possible conditions, namely, by extraction with water at room temperature. The material
thus obtained, which was further fractionated as
indicated in Fig. 12, constituted less than half of
the total pectin from this source. Varying extraction conditions were then examined in order to obtain

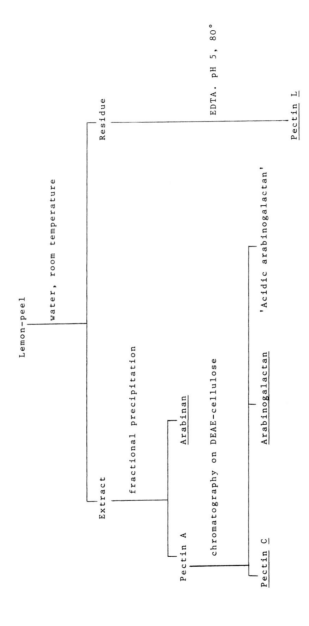

Fig. 12 Isolation and fractionation of polysaccharides from lemon-peel

the remaining pectin in undegraded form (Fig. 13).
These studies (22) showed that extraction of the
residual peel with EDTA at pH 5 led to no recog-
nizable degradation and pectin L was isolated in
this manner. We may note in passing that extraction
with sodium hexametaphosphate at pH 4.2 led to
de-esterification as well as limited degradation by
the β-elimination mechanism. The pectins C and L
obtained respectively from cold water and EDTA
extraction, were electrophoretically distinct polysaccharides, although analysis of composition indicated only rather small differences in sugar content
and in degree of esterification. Greater differences,
however, are implied if we consider the percentages
of non-esterified galacturonic acid residues, namely,
25% and 41% respectively.

In order to assess the extent to which lemon-
peel pectins C and L differed in respects other than
degree of esterification, both pectins were saponi-
fied under the mildest possible conditions, and
the corresponding pectic acids C and L were examined
(25). The results of fractionation experiments are
summarized in Fig. 14. Zitko and Bishop (26) observed that pectic acids may be fractionated by graded
precipitation with sodium acetate to give series of
polysaccharide fractions of decreasing uronic acid
and increasing neutral sugar content. The lemon-
peel pectic acids were fractionated in this way in
order to see if the slight differences in composition of the two pectic acid preparations could be
accounted for by the presence in each of two polysaccharides but in different proporations. Our
results, however, failed to provide any substantial
evidence for the presence of two such polysaccharide
species, and indeed pointed to the presence of single
essentially homogeneous polysaccharide components.
We consider that the results point to the sub-
fractionation of polydisperse systems in which each

Pectin	Method of isolation	Galac- turonic acid (%)	OMe (%)	$[\alpha]_D^\circ$	Electrophoretic mobility $(\mu \times 10)$*	
					Borate buffer	Pyridine-acetic acid buffer
A	(1) Extraction of peel with H$_2$O at room temperature	75	10.5	+206	(4.1)(6.3)11.2	5.4
C	Chromatography of pectin A	75	10.1	+212	11.3	
G	(2) Residual peel from (1) extracted with H$_2$O at pH 2.1 at 90°	80	10.3	+220		(5.4) 5.9
H	(3) Residual peel from (1) extracted with EDTA/sodium lauryl sulfate at 90°	83	10.3	+212		(5.8) 6.3
J	(4) Residual peel from (1) extracted with sodium hexametaphosphate at 90°	82	2.2	+237		12.5
K	(5) Residual peel from (1) extracted with water at 50°	83	9.1	+210	12.0	(5.4) 6.2
L	(6) Residual peel from (1) and (5) extracted with EDTA at pH 5.0 at 90°	84	8.8	+214	12.0	6.3

* Figures in parentheses indicate mobilities of minor components.

Fig. 13 Isolation of lemon-peel pectin fractions

111

Pectic acid fraction	Precipitant (M)	Galac-turonic acid(%)	$[\alpha]_D°$	Electrophoretic mobility (μ x 10^5)*		% of Parent pectic acid
				Pyridine-acetic acid buffer	Borate buffer	
C		84	+242	13.9	11.8	
C1	sodium acetate (0.12)	94	+288	14.6	n.d.	5
C2	sodium acetate (0.14)	88	+265	15.1	10.8	24
C3	sodium acetate (0.14)	85	+257	15.9	(11.6) 12.3	40
C4	sodium acetate (0.22)	82	+240	14.9	10.8	12
C5	ethanol	68	+186	15.3	11.0	8
L		87	+251	14.5	10.9	
L1a	sodium acetate (0.10)	93	+285	n.d.	11.1	5
L1b	sodium acetate (0.12)	89	+278	14.2	11.1	67
L2	sodium acetate (0.14)	84	+257	14.6	11.1	3
L3	sodium acetate (0.18)	76	+224	12.3 14.1	9.7 10.6	16

*Figures in parentheses indicate mobilities of minor components.

Fig. 14 Fractionation of lemon-peel pectic acids

contains a continuous spectrum of closely-related molecular species. Pectic acids C and L contained slightly different average proportions of galacturonic acid residues, but analyses of the subfractions indicated a substantial degree of overlap in galacturonic acid content of molecules within each spectrum of molecular species.

Our attention has been directed primarily to the structures of plant polysaccharides. A complete account of the carbohydrate polymers of cell walls must also recongize the role played by plant glycoproteins. These biopolymers are as yet much less fully characterized. The presence of the O-arabinosyl hydroxyproline linkage in these glycoproteins has been clearly established (27) and evidence is presented elsewhere in this Symposium (28) for the existence of further sugar-amino acid linkages. Several recent papers have reported the isolation of glycoprotein preparations containing not only arabinose but also galactose (29, 30) as well as rhamnose and galacturonic acid (31, 32). Since these polymers contain some or all of the common sugar constituents of pectins and associated polysaccharides a knowledge of their detailed chemical structures should provide a basis for the understanding of both biological function and metabolic inter-relationships.

References

1. J. C. Irvine and E. L. Hirst, *J. Chem. Soc.* 518 (1923); W. N. Haworth, E. L. Hirst and H. A. Thomas, *J. Chem. Soc.* 824 (1931).
2. R. H. Marchessault, and A. Sarko, *Advan. Carbohyd. Chem. Biochem.* 22, 421 (1967); D. A. Rees and R. J. Skerrett, *Carbohyd. Res.* 7, 334 (1968).
3. M. Marx-Figini, *J. Polymer Sci.* 28, 75 (1969).

4. For reviews see T. E. Timell, *Advan. Carbohyd. Chem.* 19, 247 (1964); 20, 409 (1965).

5. G. O. Aspinall, J. A. Molloy, and J. W. T. Craig, *Canad. J. Biochem.* 47, 1063 (1969).

6. W. A. Cote, Jr., A. C. Day, B. W. Simson and T. E. Timell, *Holzforschung*, 20, 178 (1966).

7. D. M. W. Anderson and I. C. M. Dea, *Phytochemistry* 8, 167 (1969) and references there cited.

8. G. O. Aspinall and C. C. Whitehead, *Canad. J. Chem.* 48, 3840, 3850 (1970).

9. G. O. Aspinall, *Advan. Carbohyd. Chem. Biochem.* 24, 333 (1969).

10. G. O. Aspinall, I. W. Cottrell, and N. K. Matheson, *Canad. J. Biochem.* 50, 574 (1972).

11. For references to earlier work see G. O. Aspinall, *Polysaccharides*, Pergamon Press, Oxford (1970).

12. V. Zitko and C. T. Bishop, *Canad. J. Chem.* 44, 1275 (1966).

13. S. S. Bhattacharjee and T. E. Timell, *Canad. J. Chem.* 43, 758 (1965).

14. D. A. Rees and A. W. Wight, *J. Chem. Soc. (B)*, 1366 (1971).

15. G. O. Aspinall and J. Baillie, *J. Chem. Soc.* 1702 (1963).

16. H. O. Bouveng, *Acta, Chem. Scand.* 19, 953 (1965).

17. G. O. Aspinall, I. W. Cottrell, S. V. Egan, I. M. Morrison, and J. N. C. Whyte, *J. Chem. Soc. (C)*, 170. (1967).

18. D. A. Rees and N. J. Wight, *Biochem. J. 115*, 431 (1969).

19. G. O. Aspinall, J. W. T. Craig, and J. L. Whyte, *Carbohyd. Res.* 7, 422 (1968).

20. G. O. Aspinall, B. Gestetner, J. W. Molloy, and M. Uddin, *J. Chem. Soc. (C)*, 2554 (1968).

21. P. Albersheim, H. Nuekom, and H. Deuel, *Arch. Biochem. Biophys*, 90, 46 (1960).

22. G. O. Aspinall and I. W. Cottrell, *Canad. J. Chem.* 48, 1283 (1970).

23. A. J. Barrett and D. H. Northcote, *Biochem. J. 94*, 617 (1965).

24. G. O. Aspinall and I. W. Cottrell, *Canad. J. Chem.* *49*, 1019 (1971).
25. G. O. Aspinall, I. W. Cottrell, J. A. Molloy, and M. Uddin, *Canad. J. Chem. 48*, 1290 (1970).
26. V. Zitko and C. T. Bishop, *Canad. J. Chem. 43*, 3206 (1965).
27. D. T. A. Lamport, *Biochemistry 8*, 1155 (1969).
28. D. T. A. Lamport, *This Symposium*, Chap. 8.
29. M. F. Heath and D. H. Northcote, *Biochem. J. 125*, 953 (1971).
30. J. A. Monro, R. W. Bailey, and D. Penny, *Phytochemistry 11*, 1597 (1972).
31. A. Pusztai and W. B. Watt, *Eur. J. Biochem. 10*, 523 (1969).
32. A. Pusztai, R. Begbie, and I. Duncan, *J. Sci. Fd. Agric. 22*, 514 (1971).

THE STRUCTURE OF THE WALL OF SUSPENSION-CULTURED SYCAMORE CELLS*

Peter Albersheim, W. Dietz Bauer,
Kenneth Keestra, and Kenneth W. Talmadge
Department of Chemistry
University of Colorado
Boulder, Colorado

Abstract

Isolated walls of cultured sycamore (*Acer pseu-doplatanus*) cells [were] [deg]raded with purified enzymes to obtain defi[...] [...] fragments were purified and [...] [st]udied using gas chromatogra[...] [...]try methylation analysis. A purified [...]acturonase solubilizes about half of th[e] [...] polymers. Treatment of the endo-polygalac[...] [...]e pretreated walls with an endoglucanase yields fragments of a xyloglucan, and, attached to these, a portion of the remaining pectic polymers. After treatment with endopolygalacturonase and endoglucanase, treatment with protease releases fragments of the structural protein of the wall as well as a portion of the remaining pectic fragments. A tentative wall structure is proposed based on the structures of the fractions solubilized by the enzyme treatments and by treatment with either alkali, urea or mild acid. In this model, xyloglucan, which is covalently attached to the pectic polymers, also binds tightly to cellulose by hydrogen bonds. This

---

*Supported in part by Atomic Energy Commission contract #AT(11-1)-1426.

results in a connection between the cellulose fibers and the pectic polymers. The pectic polymers are also covalently connected to the wall protein. This network of protein, pectic polymers and xyloglucan serves to cross-link the cellulose fibers of the wall. These macromolecules account for the entire structural component of this primary cell wall. More recent results (B. M. Wilder and P. Albersheim) have shown that the primary walls of suspension-cultured bean (*Phaesolus vulgaris*) cells have a structure extraordinarily similar to that possessed by the sycamore cells.

Introduction

The cell wall is conveniently considered to be of two types, a thin primary wall and a thicker secondary wall. The primary wall is that part of the wall laid down by young, undifferentiated cells that are still growing. The primary cell wall is transformed into a secondary wall after the cell has stopped growing. The primary cell walls of a variety of higher plants appear to have many features in common and may, in fact, have very similar structures. This is not true of secondary walls where the composition and ultrastructure vary considerably from one cell type to another. The results I will report to you come predominantly from studies of walls isolated from suspension-cultured sycamore (*Acer pseudoplatanus*) cells. More recent results indicate that walls isolated from suspension-cultured bean (*Phaseolus vulgaris*) cells are very similar to the walls of the distantly related sycamore cells.

There were several reasons for selecting the cell walls isolated from suspension-cultured sycamore cells for the present studies. The most important is that cultured sycamore cells are a homogeneous tissue possessing primary but no secondary walls. A second attribute of cultured cells is that they secrete into their culture medium polysaccharides

which are similar in composition to the non-cellulosic
polysaccharides of the cell wall (14). It has been
suggested (14) and now confirmed (13) that at least
some of these extracellular polysaccharides are
structurally related to cell wall polysaccharides.
Therefore, these extracellular polymers offer a
convenient source of material for developing
techniques to study the wall polymers.

The sugar composition of the total wall of isolated polysaccharides, or of polysaccharide fragments has been obtained by the formation of alditol
acetate derivatives (Fig. 1). These volatile derivatives may be separated and quantitatively analyzed by
gas chromatography (Fig. 2) and computer assisted
data reduction of the gas chromatographic output.
This method allows relatively quick, facile, and
accurate analysis of sub-milligram quantities of all
of the neutral and acidic sugars present in plant
cell walls (24).

The structural analysis of cell walls has been
facilitated by recent improvements in the technique
of methylation analysis. The complete methylation
of most polysaccharides can now be achieved in one
or at most two methylation reactions by the method
of Hakomori (23). Probably the most important ad-
vance in the area of methylation analysis has been
the development of an easy procedure for the qualitative and quantitative analysis of the methylated
sugars obtained upon acid hydrolysis of methylated
polysaccharides (15). In this procedure, the partially methylated sugars obtained by hydrolysis of
the permethylated polysaccharides are converted to
the O-methyl alditols; these are then acetylated
(Fig. 3). The resulting partially methylated alditol
acetates are separated, identified, and quantitatively
estimated by the combined application of gas chromatography and mass spectrometry. The mass spectrometric identifications are based on the fact that
a unique fragmentation pattern is obtained for each

119

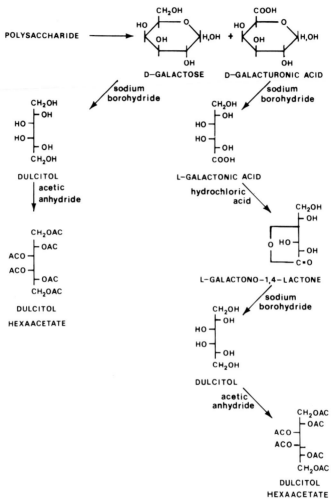

Fig. 1 Preparation of alditol acetate derivatives of sugars and uronic acids for gas chromatography. Cell wall polysaccharides are depolymerized by mild acid hydrolysis followed by treatment with a mixture of polysaccharide-degrading enzymes. The aldoses and hexuronic acids which are liberated are then reduced to alditols and aldonic acids, respectively. These two classes of compounds are separated by means of Dowex-1 anion exchange resin. The alditols and aldonic acids are then separately converted into alditol acetates by the indicated steps (24).

120

substitution pattern (position of O-methyl groups on the sugar alcohol) in partially methylated alditol acetates. These mass spectrometric data, in combination with relative retention time data obtained from several gas chromatographic columns, will, in most cases, give sufficient evidence for an unambiguous identification of the methylated sugar. The major advantage of this procedure is that crystalline derivatives are not necessary for the identification of the methylated sugars; a methylation analysis can be performed accurately on milligram quantities of material.

Although the general components of plant cell walls have been identified and studied in considerable detail, no clear picture of the molecular structure of any plant cell wall has been presented. Nor could such a picture be drawn until the general components of the wall were resolved into distinct polymers of definite composition, linkage, size and sequence, and until the interconnections between these polymers were identified (13, 25, 37). The major difficulty in any analysis of cell wall structure has been finding suitable methods for the isolation of *defined* fragments from the wall. Classical methods of extraction, using acid or base, resulted in the simultaneous, but partial, cleavage of several types of bonds present in the plant cell wall. This lack of specificity made it most difficult to determine how the fragments extracted with acid or base were linked to each other or to the residual wall structure. A much more satisfactory fragmentation of the cell wall has been achieved through the use of purified hydrolytic enzymes. The value of this method was clearly demonstrated by the many elegant studies of bacterial wall structure (21). The specificity of a given hydrolytic enzyme can be determined with model substrates, and, with this knowledge, the linkages between enzymically released wall fragments and the residual wall may be reconstructed.

123

Those plant pathogens which have the capacity
to degrade the cell walls of their hosts represent
an excellent source of the highly specific polysaccharide-degrading enzymes required for the structural analysis of cell walls (3, 12). The fungal
pathogen *Colletotrichum lindemuthianum* has been shown
to secrete different enzymes in a temporal sequence
with regard to culture age (19). The first enzyme
secreted by this fungus is an endopolygalacturonase.
This enzyme has been highly purified (20) and the
purified enzyme is able to remove most of the
galacturonic acid from the isolated cell walls of a
number of plants. This purified endopolygalacturonase
has been very important in our present studies as it
is the only enzyme which, by itself, has been found
to initiate effectively the degradation of isolated
cell walls.

The Cell Wall Hemicellulose

For an example of the efficacy of purified
polysaccharide-degrading enzymes when combined with
gas chromatographic-mass spectrometric analysis, let
us examine briefly the structure of a xyloglucan, the
only hemicellulose present in suspension-cultured
sycamore walls. The gas chromatogram in the top of
Fig. 4 shows the partially methylated alditol acetates derived from this polymer (13). The inositol
in this chromatogram has been added as an internal
standard. Some of the peaks contain more than a
single component; these have been separated by gas
chromatography on a different column. Each of the
components in the gas chromatogram, including the
arabinose, fucose and galactose derivatives, are
part of the polymer which we call xyloglucan.

A truly beautiful example of conservation of
structure during evolution was obtained by Barry
Wilder in my laboratory when he compared the gas
chromatograms of the partially methylated alditol
acetates obtained from the purified xyloglucan of

124

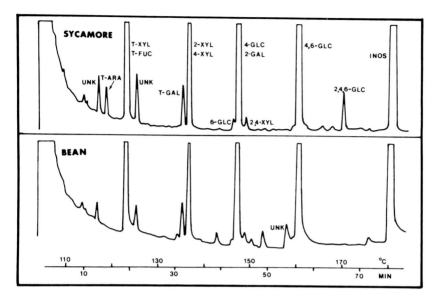

Fig. 4 Gas chromatograms of the methylated alditol acetates obtained from purified xyloglucan of
sycamore (top) (13) and of bean (bottom) (Wilder and
Albersheim, unpublished results). The initial peak
in the chromatograms is due to the acetic anhydride
used as the solvent. The glycosidic linkages to
each sugar derivative are indicated by numerical
prefixes: thus, 4,6-GLC indicates that sugars are
glycosidically linked in the polysaccharide to the
4 and the 6 carbons of glucosyl residues. Terminal
residues are indicated by T-(e.g., T-XYL). The
sugars are arabinose (= ARA), fucose (= FUC), galactose (= GAL), glucose (= GLC), xylose (= XYL), and
myo-inositol (= INOS) which was used as a standard.
Unidentified components are indicated (= UNK).

suspension cultured bean cells (Fig. 4, bottom) with
the chromatogram of the xyloglucan of the very distantly related sycamore cells (Fig. 4, top). This
comparison, in addition to many other similarities,
suggests that the primary cell walls of all flowering
plants are likely to be very similar.

125

The detailed molecular structure of the sycamore xyloglucan, as well as of bean xyloglucan, has been characterized by the methylation analysis of the oligosaccharides obtained by endoglucanase treatment of the polymer. This enzyme hydrolyzes the glycosidic bonds of glucosyl residues that have another sugar attached to carbon 4, but the enzyme does not hydrolyze the glycosidic bonds of glycosyl residues that have other sugars attached to both carbons 4 and 6. The fragments produced by endoglucanase treatment were separated by chromatography on Bio-Gel P-2 (Fig. 5). Each fragment has been structurally analyzed (the analyses of peaks 3 and 4 are given below) and the structure of the polymer was, in large part, deduced from this information (13). The structure of the xyloglucan is based on a repeating heptasaccharide unit which consists of 4 residues of β-1,4--linked glucose and 3 residues of terminal xylose. A single xylose residue is glycosidically linked to carbon 6 of three of the glucosyl residues. Figs. 6 and 7 illustrate the type of data obtained to characterize each of these xyloglucan fragments. The xyloglucan appears to consist predominantly of repeating units of the seven sugar fragment of Fig. 6 and the nine sugar fragment of Fig. 7. This latter fragment contains, in addition to the basic heptose repeating unit of Fig. 6, a fucosylgalactose disaccharide attached to one of the xylose residues.

The function of the xyloglucan appears to be based on the ability of its β-1,4-linked glucan backbone to hydrogen bond to the β-1,4-glucan chains of cellulose fibrils. Models of the xyloglucan indicate that fucosylgalactose side chains prevent further β-1,4-glucan chains from hydrogen bonding to those xyloglucan chains which are already hydrogen-bonded to cellulose fibrils. This results in the cellulose fibrils being coated with a single layer of xyloglucan chains. It appears that there is sufficient xyloglucan in suspension-cultures sycamore cell walls

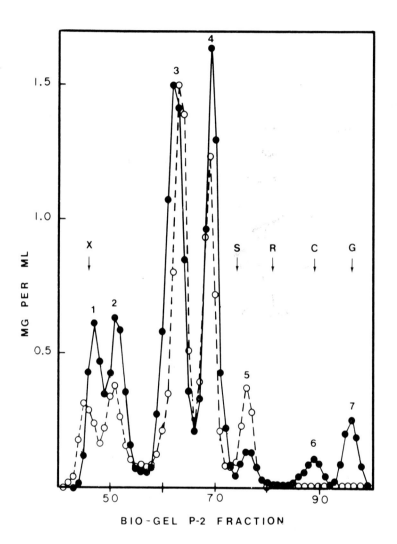

Fig. 5 Bio-Gel P-2 fractionation of endoglu-canase-treated sycamore xyloglucan. Untreated xylo-glucan voids the column. The elution volumes of untreated xyloglucan (X), stachyose (S), raffinose (R), cellobiose (C), and glucose (G) are indicated by arrows.

COMPOSITION		LINKAGE	COMPOSITION	
EXPT	CALC		EXPT	CALC
XYL 44.1	42.9	T—XYL	41.3	42.9
GLC 55.9	57.1	4—GLC	12.7	14.3
		6—GLC	13.4	14.3
		4,6—GLC	27.3	28.6

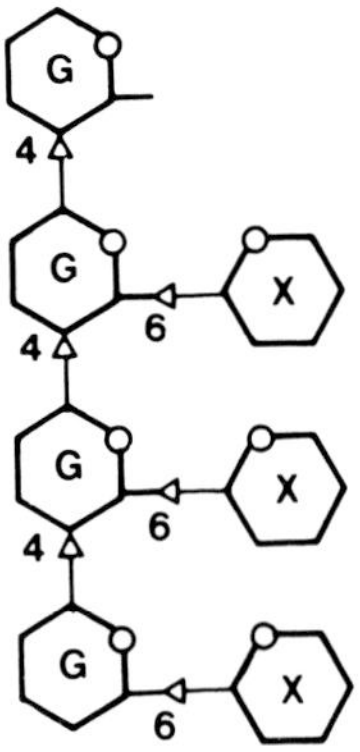

Fig. 6 The structure of the endoglucanase-derived xyloglucan oligosaccharide which elutes as peak 4 from the Bio-Gel P-2 column depicted in Fig. 5 (13). The alditol acetate-determined sugar composition and the partially methylated alditol acetate-determined linkage composition are compared to the values anticipated for the proposed structure. The sugars are glucose (= GLC or G) and xylose (= XYL or X).

	COMPOSITION		LINKAGE	COMPOSITION	
	EXPT	CALC		EXPT	CALC
FUC	10.3	11.1	T—FUC	11.1	11.1
XYL	33.0	33.3	T—XYL	20.4	22.2
GAL	12.1	11.1	2—XYL	11.3	11.1
GLC	45.2	44.4	T—GAL	2.3	0
			2—GAL	9.4	11.1
			4—GLC	12.8	11.1
			6—GLC	10.8	11.1
			4,6—GLC	21.1	22.2

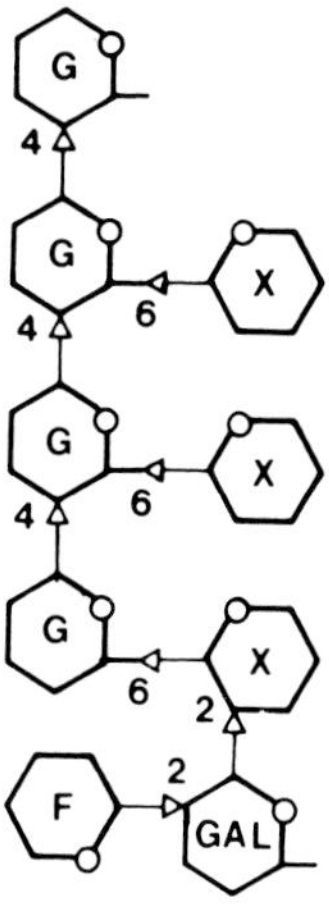

Fig. 7 The structure of the endopolyglucanase-
derived xyloglucan oligosaccharide which elutes as
peak 3 from the Bio-Gel P-2 column as depicted in
Fig. 5 (13). The alditol acetate-determined sugar
composition and the partially methylated alditol
acetate-determined linkage composition are compared
to the values anticipated for the proposed structure.
The particular xylose residue to which the fucoxyl-
galactose is attached is unknown. The sugars are
fucose (= FUC or F), galactose (= GAL), glucose
(= GLC or G), and xylose (= XYL or X).

to coat completely all of the elementary cellulose
fibrils. The reducing end of the xyloglucan chains

is covalently attached to the pectic polymers. Thus,
the xyloglucan chains serve as a bridge between each
cellulose fibril and the rest of the cell wall.

The Cell Wall Pectic Polymers

The walls of suspension-cultured sycamore cells
contain a single acidic pectic polymer, a rhamnogalacturonan, and three neutral pectic polymers, a
4-linked galactan, a highly branched arabinan, and
a 3,6-linked arabinogalactan. A tentative structure
for the rhamnogalacturonan (the acidic pectic polymer)
of sycamore walls is presented in Fig. 8. This
structure is based on the analyses of fractions obtained by endopolygalacturonase treatment of cell
walls and on the analysis of the galacturonosyl-con-
taining oligomers obtained by partial acid hydrolysis
of isolated cell walls. The rhamnogalacturonan is
not a straight chain molecule. The zigzagged shape
results from the presence of 2-linked rhamnosyl
residues in an otherwise α-(1→4)-linked galacturonan
chain. When another sugar is attached to carbon 4
of a rhamnosyl residue, the rhamnose forms a Y-shaped
branch point. These observations came as a result of
building CPK space filling models of these structures.
A similar observation was made by Simmons in studying
the O-antigen polysaccharides of *Shigella flexneri*.
He noted that the α-(1→2)-rhamnosyl linkages of that
oligosaccharide caused a "buckling" of the otherwise
linear polymer (34).

The rhamnosyl residues are not randomly distributed in the chain, but probably occur as rhamnosyl-(1→4)-
galacturonosyl-(1→2)-rhamnosyl units. This sequence
appears to alternate in the polymer with a homogalacturonan sequence containing approximately 8 residues
of 4-linked galacturonic acid. These two sequences
give rise to the two major fractions obtained from
endopolygalacturonase treatment of sycamore cell walls.
Mono-, di-, and tri-galacturonic acid arise from the
endopolygalacturonase hydrolysis of the homogalacturonan

sequences of the rhamnogalacturonan polymer, whereas
neutral sugar-rich polymers arise from the rhamnose
rich region of the rhamnogalacturonan polymer.

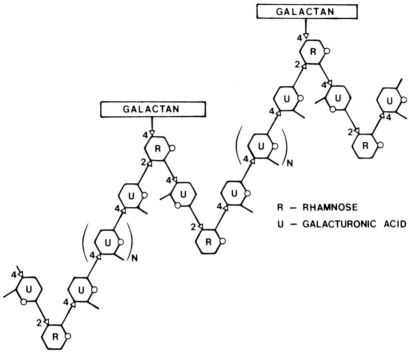

Fig. 8 A proposed structure for the rhamnogalacturonan of sycamore cell walls. The structure is
based on evidence presented by Talmadge et. al (37).
The number of residues in the homogalacturonan regions is unknown, but "N" is estimated to be about 6.

Degradation of the galacturonan chain by the
endopolygalacturonase demonstrates that the glycosidic
linkages of the galacturonosyl residues are in the
alpha configuration (20). Diborane reduction results
of methylated di- and tri-galacturonosides substan-
tiate that the galacturonosyl residues in the galacturonan chains are 4-linked. The linear galacturonan
portion of the rhamnogalacturonan (Fig. 8) gives rise
to larger oligo-galacturonosides when intact cell
walls are subjected to partial acid hydrolysis. These

131

oligo-galacturonosides, which represent about 5% of
the cell wall, have rhamnosyl residues at their
reducing ends. The presence of rhamnose in these
oligomers provides evidence that the two regions in
the proposed structure are linked in an alternating
sequence in the same polymer. The estimate of the
number of galacturonosyl residues in the galacturonan region was obtained from gel filtration chroma-
graphy of the oligo-galacturonosides isolated following partial acid hydrolysis of cell walls.

Additional evidence for the presence of rhamnosyl residues in the galacturonan chain was provided by the finding that the most abundant aldouronic acid present in acid-hydrolyzed cell walls is
galacturonosyl-$(1\rightarrow2)$-rhamnose. This same aldouronic
acid has also been reported to occur in the pectic
polysaccharides obtained from a number of different
plant sources, including sycamore cell walls (8, 9,
11, 32, 36). The presence of terminal rhamnosyl
residues in the endopolygalacturonase products but
not in the unfractionated cell wall is evidence that
the endopolygalacturonase, in addition to breaking
the bonds between adjacent 4-linked galacturonosyl
residues, also cleaves a signicant percentage of the
$(1\rightarrow2)$-linked glycosyl bonds between galacturonosyl
and rhamnosyl residues. This observation constitutes
still additional evidence that galacturonosyl residues are linked to rhamnosyl residues.

As indicated previously, the rhamnose-rich
sequence in Fig. 8 gives rise to the neutral sugar-
rich polymers when cell walls are treated with
endopolygalacturonase. These neutral sugar-rich polymers are not further degraded by the endopolygalacturonase, demonstrating that there are no more than
two galacturonosyl residues between the rhamnosyl
residues. The fact that the predominant aldouronide
isolated from partially acid hydrolyzed cell walls
is the disaccharide galacturonosyl-$(1\rightarrow2)$-rhamnose

132

rather than the trisaccharide galacturonosyl-(1→4)-galacturonosyl(1→2)-rhamnose indicates that in the neutral sugar-rich region there is only one galacturonosyl residue between neighboring rhamnosyl residues. These results, along with the fact that the neutral sugar-rich region contains galacturonic acid and rhamnose in the molar ratios of 2:1, suggest that the rhamnogalacturonan portion of the neutral rich polymers has an average structure of galacturonosyl-(1→2)-rhamnosyl-(1→4)-galacturonosyl-(1→2)rhamnosyl-(1→4)-galacturonosyl-(1→4)-galacturonosyl.

DEAE-Sephadex ion exchange chromatography demonstrates that the neutral sugars of the endopolygalacturonase products are covalently attached to galacturonosyl residues. Most of these neutral sugar residues, which represent 76% of this fraction, are part of either a branched arabinan or a linear 4-linked galactan. Methylation analysis has shown that approximately 50% of the rhamnosyl residues in the wall are branched, having a substituent at carbon 4 as well as a galacturonosyl residue attached to carbon 2. As this is the major branch point of the rhamnogalacturonan chain, the 2,4-linked rhamnosyl residues are likely to represent the point of attachment of the 4-linked galactan chains. The point of attachment of the branched arabinan chains is not known.

Most of the arabinosyl and galactosyl residues that are released by endopolygalacturonase treatment are part of acidic polymers. The structure of these polymers was examined by partial acid hydrolysis. Following such hydrolysis, the sample was fractionated on DEAE-Sephadex into neutral and acidic components which were then separately chromatographed on a Bio-Gel P-2 column. The results from this analysis indicate that about 85% of the arabinosyl residues were released by hydrolysis from the acidic rhamno-

galacturonan fragment while 75% of the galactosyl
residues remain attached to the acidic fragment.
This preferential cleavage of arabinose is accompanied by only minor changes in the linkages of
the sugar residues remaining in the acidic fraction.
These results, in conjunction with methylation
analysis which showed the presence of only small
amounts of branched galactosyl residues and large
amounts of terminal and branched arabinosyl residues,
suggest that the galactosyl residues are present as
a linear chain which is attached at its reducing end
to the rhamnosyl residues of the rhamnogalacturonan
main chain, and that the arabinosyl residues are in
the form of a branched chain. Recent results
(McNeil and Albersheim, unpublished) suggest that
the arabinan is attached to the rhamnogalacturonan
rather than to the galactan.

The endopolygalacturonase products contain, in
addition to the arabinan, the 4-linked galactan, the
rhamnogalacturonan fragment, and small amounts of
xyloglucan. The xyloglucan of the endopolygalacturo-
nase-liberated polymers fractionates as an acidic
polymer on DEAE-Sephadex, indicating that there is a
covalent attachment between the neutral xyloglucan
chains and the acid pectic polysaccharides. After
weak acid hydrolysis of this portion of the cell
wall, over 70% of the xyloglucan still fractionates
as an acid polymer on DEAE-Sephadex. This indicates
that the xyloglucan may be attached to the rhamnogalacturonan via the galactan; and this result is
further evidence that acid-labile arabinosyl residues
are not interspersed in the galactan chain. Thus,
the galactan may serve as a cross-link between the
xyloglucan and rhamnogalacturonan components of the
wall.

The structures proposed for the sycamore pectic
polymers are generally consistent with previous
structural studies on pectic polysaccharides. Rhamno-
galacturonans appear to be a common feature of all

pectic polysaccharides. Pectic galactans containing a β-1,4-linked galactan backbone have been isolated from soybean seed (6). The highly branched arabinan region of sycamore pectic polysaccharides contain similar linkages to those present in the pectic arabinans isolated from soybean (7), lemon peel (7), and mustard seed (7).

Another wall polysaccharide has been identified which contains both arabinosyl and galactosyl residues. This arabinogalactan is only questionable listed as a pectic polymer. Our evidence suggests that this arabinogalactan may function as a covalent link between the rhamnogalacturonan chains and the hydroxyproline-rich wall protein. Most of our work with this polymer has been carried out with the arabinogalactan isolated from the extracellular polysaccharides of suspension-cultured sycamore cells (unpublished results).

The arabinogalactan, at pH 2, absorbs to a SE-Sephadex column and, when such a column is eluted with a linear salt gradient, the arabinogalactan elutes simultaneously with a hydroxyproline-containing protein. This suggests that the arabinogalactan and the protein are parts of the same molecule.

Methylation analysis indicates that the arabinogalactan possesses a highly branched structure containing predominantly 3,6-linked galactosyl residues as branch points with single arabinosyl residues as the most prevalent side chain. These results are similar to those reported by Aspinall et al. (10) for this arabinogalactan and are also similar to arabinogalactans isolated from coniferous woods (38) and from plant gums (4). The sycamore extracellular arabinogalactan differs from those in wood in that it contains a rhamnosyl residue and a higher percentage of arabinose. An arabinogalactan isolated from maple (*Acer saccharum*) is similar to the sycamore arabinogalactan in all respects (1).

One possible structure for the arabinogalactan from sycamore is shown in Fig. 9. This structure has two features which suggest that it is an interesting cell wall component. The first is the attachment of this polysaccharide to a hydroxyproline-containing protein, a known component of plant cell walls (10, 26). This suggests that this arabinogalactan may be a wall component, and that it may be a connecting point between wall polysaccharides and wall protein. Recent evidence from Lamport's laboratory as well as from our laboratory suggests that the arabinogalactan may be attached to the hydroxyproline-rich protein by linkage to the hydroxyl groups of serine residues (personal communication). The second striking feature of the arabinogalactan is the presence of a terminal rhamnosyl residue. Since all of the rhamnose in the cell wall is thought to be covalently linked in the rhamnogalacturonan (37), the rhamnosyl residue in the arabinogalactan might act as a primer to which a rhamnogalacturonan chain is attached.

A Tentative Molecular Structure of Sycamore Cell Walls

Direct methylation analysis of the isolated but unfractionated sycamore cell wall has been important in our studies for it yielded a quantitative summary of all but the minor sugar linkages present in the total wall (37). Subsequent studies on wall fractions obtained by the action of purified hydrolytic enzymes were made quantitative by comparing the amounts of the sugar linkages found in each fraction to the amounts present in the total wall. These results demonstrate that the sycamore cell wall is composed of a limited number of major structural components: a branched arabinan, cellulose, a 4-linked galactan, the hydroxyproline-rich glycoprotein, rhamnogalacturonan, xyloglucan, and a small

136

SUGAR LINKAGE	MOLAR RATIO
TERM ARA	8
2,5 ARA	2
TERM GAL	1
3 GAL	2
6 GAL	2
3,6 GAL	7
TERM RHA	1
URONIC ACID	2

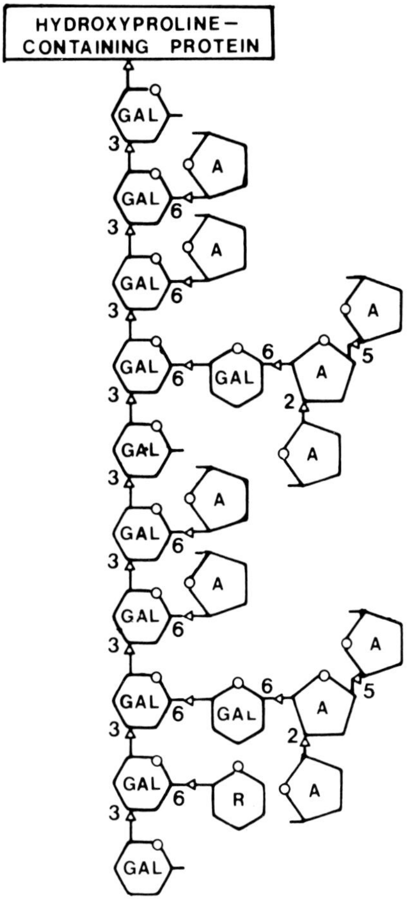

A = ARABINOSE

GAL = GALACTOSE

R = RHAMNOSE

Fig. 9 The molar ratio of the glycosyl derivatives present in sycamore extracellular arabinogalactan and a proposed structure for this polysaccharide. The molar ratios were determined by Keegstra et. al. (25). Since uronic acids are not recovered from the methylation analysis under the conditions used, it was not possible to tell how the uronic acids are linked in this structure. The structure shown is not unique to the data, but it is consistent with and accounts for the data available. This polymer may be connected to the serine residues of the hydroxyproline-rich structural wall protein.

amount of 3,6-linked arabinogalactan (Table 1).

Table 1

Polymer Composition of Suspension-Cultured Sycamore Cell Walls

Wall Component	Cell Walls %
Arabinan	10
3,6-linked arabinogalactan	2
4-linked galactan	8
Cellulose	23
Protein	10
Rhamnogalacturonan	16
Tetra-arabinosides (attached to hydroxyproline)	9
Xyloglucan	21
Total	99

Essentially the entire cell wall is accounted for by the sum of the amounts of these seven polymeric components. The impression we had when we started this study was that the plant cell wall was made up of a relatively large number of complex polymers. The fact that the sycamore cell wall is composed of a limited number of well-defined structural components is a finding of utmost importance to plant cell wall research. Future studies of other plant cell walls and of the structural changes which take place during growth and development will be encouraged and facilitated by this knowledge.

A model (Fig. 10) of the sycamore cell wall can be proposed from our results (25). Although other structures are possible, the one presented is consistent with all of the data obtained. The model utilizes the fact that xyloglucan has been shown to

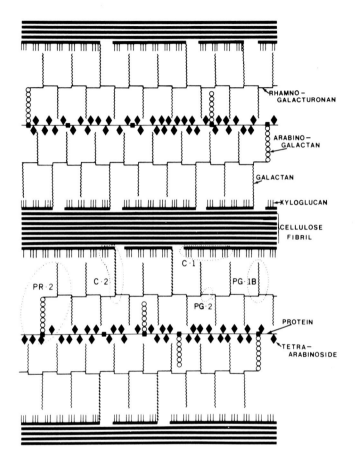

Fig. 10 A tentative structure of the walls of suspension-cultured sycamore cells (25). The structure presented is discussed in the text. This model is not intended to be quantitative, but an effort has been made to present the wall components in approximately proper proportions. The distance between cellulose elementary fibrils is expanded to allow room to present the interconnecting structure. There are probably between 10 and 100 cellulose elementary fibrils across a single primary cell wall.

bind tightly to purified cellulose (4) as well as to
the cellulose of the cell wall (13). Several lines
of experimental evidence (25) indicate that the reducing ends of the xyloglucans are attached to the
galactan side chains of the rhamnogalacturonan (Fig.
10). A single pectic polysaccharide is likely to
be attached through xyloglucan chains to more than
one cellulose fibril; and a single cellulose fibril
attached through xyloglucan chains to more than one
pectic polysaccharide. Such an arrangement would
result in cross-linking of the cellulose fibrils.
The hypothetical structure presented in Fig. 10 suggests, too, that the pectic polysaccharides are attached to seryl residues of the wall protein through
a 3,6-linked arabinogalactan chain (25). This connection would also result in cross-linking of the
cellulose fibrils. The wall is likely to possess
both modes of cross-linking.

The structure presented in Fig. 10 allows an
understanding of the wall fragments that each enzyme
releases. An example of that portion of the wall
solubilized by each enzyme is circled and labeled.
These fractions are summarized as follows. The
endopolygalacturonase attacks the galacturonosyl
linkages of the main pectic chain releasing tri-,
di-, and monogalacturonic acid (Fraction PG-2) as
well as arabinan and galactan side chains attached
to acidic fragments of the main rhamnogalacturonan
chain (Fraction PG-1B). After the pectic polysaccharide has been partially degraded by the endopolygalacturonase, endoglucanse more readily degrades
xyloglucan releasing neutral oligosaccharides (Fraction C-1) as well as pectic fragments that had been
held insoluble by their connection with xyloglucan
(Fraction C-2). Pronase, which cannot release significant amounts of carbohydrate from untreated walls,
is able to release pectic fragments not released
by endopolygalacturonase pretreatment, and larger
amounts of carbohydrate are released by pronase
after a combination of endopolygalacturonase and

140

endoglucanase treatment (Fraction PR-2).

The primary cell wall of sycamore cells can be considered as a single macromolecule (Fig. 10). The components of the cell wall are, with the exception of the connection between xyloglucan and cellulose, interconnected by covalent bonds; and the many hydrogen bonds which interconnect cellulose and xyloglucan make this connection as strong as a covalent bond. It has been suggested (28) that the plant cell wall contains a protein-glycan network analogous to the peptido-glycan network of bacterial cell walls (21). Our results support this analogy as we find that the structural component of the sycamore cell wall is composed of well-defined interconnected polymers in the form of a large "bagshaped" molecule, a description made popular by Weidel and Pelzer (39).

The goal of many workers in cell wall research is an elucidation of the mechanism underlying control of cell wall extension. Cleland (17) has lucidly summarized the current thinking about wall extension. It is generally agreed that addition of auxin to tissues deficient in this hormone quickly causes the primary cell walls of the tissue to be loosened or weakened, such that the rate of cell extension is increased. Perhaps not as widely held, but nevertheless accepted by us, is the view that auxin initiates wall loosening so quickly that *de novo* protein synthesis and *de novo* polysaccharide synthesis cannot participate in this initiation. Thus, we have examined our structural model of the primary cell wall with the idea that initiation of wall extension probably results from a rearrangement or alteration of the existing wall structure. Regardless of the mechanism involved, since walls grow throughout their length, in order for a wall to elongate, the cellulose fibrils within the wall must be able to slide along their length relative to each other.

In the walls of suspension-cultured sycamore cells (Fig. 10), there are several polymer connections whose cleavage could result in the weakening of the wall. For example, hydrolysis of the connection between the rhamnogalacturonan and the hydroxy-proline-rich protein might effectively loosen the wall. On the other hand, enzymic hydrolysis of cellulose, xyloglucan, the 4-linked galactan, or the rhamnogalacturonan would have to quite extensive in order to loosen the wall significantly. Further examination of the wall structure suggests an attractive alternative to a hydrolytic cleavage mechanism. The only non-covalent cross-link between the structural polymers of the wall is the hydrogen bond-mediated connection between the xyloglucan chains and the cellulose fibers. Extension of the wall would result if the xyloglucan chains and the cellulose fibers moved relative to one another.

Movement of the xyloglucan chains along the cellulose fibers could be accomplished by a non-enzymatically catalyzed creep, that is, by the xyloglucans moving like inch worms along the cellulose fibers. If such a mechanism exists, the rate at which the xyloglucan chains move along the surface of the cellulose fibrils would be increased by conditions which weaken hydrogen bonds, conditions such as high hydrogen ion concentrations and elevated temperatures. The conditions which would control the rate of xyloglucan creep are exactly those that Cleland (17) concludes are able to regulate the rate limiting step of cell wall extension. Furthermore, xyloglucan creep suggests that auxin's ability to weaken the wall, and thereby, to stimulate growth could result from an auxin-activated hydrogen ion pump in the cell membrane. Auxin might act by lowering the pH of the cell wall and thereby enhancing the rate of xyloglucan creep.

142

Are the Features of the Sycamore Cell Wall Structure Found Universally in the Primary Cell Walls of All Plants?

Another important aspect of the model presented is that it provides a framework for interpreting results already obtained. It is rather difficult to compare the structures of the wall components described here with the data in the literature because of the wide variety of wall preparatory procedures used and because of the heterogeneity of chemically extracted fractions (18, 31, 36). Most of the preparatory procedures that have been used result in the presence of water soluble polymers in the wall preparations. While these polymers are interesting in their own right, their presence confuses the study of the structural portion of the wall. In addition, the chemical extraction procedures which have been used to extract the pectic polymers (28) result in the hydrolysis of such bonds as arabinosyl or rhamnosyl glycosides (37). On the other hand, the strong alkali used to extract hemicellulose simultaneously results in transelimination of uronic acids (2, 30) and β-elimination of serine glycosides (35).

Despite these difficulties, there are important findings that have been reported in the literature which are consistent with the results reported here. Lamport (27) has used chemical extraction procedures to determine that cell walls isolated from suspension-cultured sycamore cells contain 36% pectin, 34% hemicellulose and 27% cellulose. He reports that most of the wall protein is in the hemicellulose fraction. Roelofsen (33) has noted that primary cell walls are typically one-third cellulose, one-third hemicellulose, and one-third pectin plus protein. Values calculated from our results (25), including both protein and xyloglucan in the hemicellulose fraction, are 34% pectin, 38% hemicellulose, and 26% cellulose. Thus, our values are in good agreement

with the typical primary wall values given by
Roelofsen as well as with the sycamore wall values
given by Lamport.

Many of the specific features of the sycamore
cell wall are found in cell walls of other plants.
The hydroxyproline-rich protein with its associated
oligo-arabinosides are widespread in the plant kingdom (28, 29). Kooiman (26) has demonstrated that
xyloglucan is found in the cell walls of the cotyledons or endosperm of a wide variety of plants.
Wilder and Albersheim (unpublished results) have
found xyloglucan in the walls of suspension-cultured
bean cells. There is even a report which provides
some evidence of a connection between the xyloglucan
and pectic polysaccharides of the cell walls of
mustard cotyledons; the report describes a pectic
polysaccharide which has been purified to a state
that "if not homogeneous, consists of a family of
related species" (32). Methylation analysis was
used to demonstrate that this preparation contained
xyloglucan as well as the pectic polymers. Although the authors considered the xyloglucan to be
a contaminant, we interpret their data as evidence
in support of a covalent linkage between these wall
components.

There may be variations in the polysaccharides
which constitute the structural portion of the cell
wall. For example, it is known that several hemicelluloses other than xyloglucan bind tightly to
cellulose (16). These hemicelluloses may, in some
plants, substitute for xyloglucans as a connector
between cellulose and the pectic polysaccharides.
Variation has been observed in the arabinose and
galactose composition of the pectic polysaccharides
(5). Although the chains containing these sugars may
vary in composition and detailed structure, these
polysaccharides can still serve to bridge co-valently
the hemicellulose and polyuronide chains, as well as
the polyuronide chains and the structural protein.

An interesting observation concerning the structure of plant cell walls has been reported by Grant et al. (23). They have isolated a soluble mucilage particle from mustard seedlings and have speculated that this particle may represent a structural unit of the cell wall. The particle consists of a cellulose elementary fibril encapsulated by other polysaccharides. The composition of the encapsulating polysaccharides suggests that they are xyloglucan and pectic polymers. Thus, the "cell wall unit" of mustard seedlings may be similar to the structure of the cell walls of the distantly related sycamore tree.

The evidence available in the literature and that reported here strongly sustains the hypothesis that the interrelationship between the structural components of the primary cell walls of all higher plants is comparable.

References

1. G. A. Adams and C. T. Bishop, *Can. J. Chem. 38*, 2380 (1960).
2. P. Albersheim, *Biochem. Biophys. Res. Commun. 1*, 253 (1959).
3. P. Albersheim, T. M. Jones and P. D. English, *Ann Rev. Phytopath. 7*, 171 (1969).
4. G. O. Aspinall, *Advan. Carbohyd. Chem. 24*, 333 (1969).
5. G. O. Aspinnall in *The Carbohydrates*, W. Pigman and D. Horton, Eds. IIb, Academic Press, New York, 515.
6. G. O. Aspinall, R. Begbie, A. Hamilton and J. N. C. Whyte, *J. Chem. Soc. (C)*: 1065 (1967).
7. G. O. Aspinall, and I. W. Cottrell, *Can. J. Chem. 49*, 1019 (1971).
8. G. O. Aspinall, I. W. Cottrell, S. V. Egan, I. M. Morrison and J. N. C. Whyte, *J. Chem. Soc. (C)*, 1071 (1967).

9. G. O. Aspinall, B. Gestetner, J. A. Molloy, and M. Uddin, *J. Chem. Soc.* (C), 2554 (1968).

10. G. O. Aspinall, J. A. Molloy and J. W. T. Craig, *Can. J. Biochem.* 47, 1063 (1969).

11. A. J. Barrett and D. H. Northcote, *Biochem. J.* 94, 617 (1965).

12. D. F. Bateman, H. D. Van Etten, P. D. English, D. J. Nevins and P. Albersheim, *Plant Physiol.* 44, 641 (1969).

13. W. D. Bauer, K. W. Talmadge, K. Keegstra and P. Albersheim, *Plant Physiol.*, Accepted for publication (January, 1973).

14. G. E. Becker, P. A. Hui and P. Albersheim, *Plant Physiol.* 39, 913 ((1964).

15. H. Björndal, C. G. Hellerquist, B. Lindberg and S. Svensson, *Angew. Chem. Int. Ed. Engl.* 9, 610 (1970).

16. J. D. Blake and G. N. Richards *Carbohyd. Res.* 17, 253 (1971).

17. R. Cleland, *Ann. Rev. Plant. Physiol.* 222, 197 (1971).

18. J. E. Dever, R. S. Bandurski, and A. Kivilaan, *Plant Physiol.* 43, 50 (1968).

19. P. D. English and P. Albersheim, *Plant Physiol.* 44, 217 (1969).

20. P. D. English, A. Maglothin, K. Keegstra and P. Albersheim, *Plant Physiol.* 49, 293 (1972).

21. J. M. Ghuysen, *Bacteriol. Rev.* 32, 425 (1968).

22. G. T. Grant, C. McNab, D. A. Rees and R. J. Skerrett, *Chem. Commun.* 805 (1969).

23. S. Hakomori, *J. Biochem.* (Tokyo) 55, 205 (1964).

24. T. M. Jones and P. Albersheim, *Plant Physiol.* 49, 926 (1972).

25. K. Keegstra K. W. Talmadge, W. D. Bauer and P. Albersheim, *Plant Plysiol.*, accepted for publication (January, 1973).

26. P. Kooiman, *Acta Bot. Neerl.* 9, 208 (1960).

27. D. T. A. Lamport, *Adv. Botan. Res.* 2, 151 (1965).

28. D. T. A. Lamport, *Ann. Rev. Plant Physiol.* 21, 235 (1970).

29. D. T. A. Lamport and D. H. Miller, *Plant Physiol.*

48, 454 (1971).

30.   H. Neukom and H. Deuel, *Chem. and Ind.* 683 (1958).

31.   P. M. Ray, *Biochem. J. 89*, 144 (1963).

32.   D. A. Rees and N. J. Wight, *Biochem. J. 115*, 431 (1969).

33.   P. A. Roelofsen, in *The Plant Cell Wall*.. Borntraeger, Berlin p. 128 (1959).

34.   D. A. R. Simmons, *Eur. J. Biochem. 18*, 53 (1971).

35.   R. G. Spiro, *Ann. Rev. Biochem. 39*, 599 (1970).

36.   R. W. Stoddart, A. J. Barrett and D. H. Northcote, *Biochem. J. 102*, 194 (1967).

37.   K. W. Talmadge, K. Keegstra, W. D. Bauer and P. Albersheim, *Plant Physiol*. accepted for publication (January, 1973).

38.   T. E. Timell, *Advan. Carbohyd. Chem. 20*, 409 (1965).

39.   W. Weidel and H. Pelzer, *Advan. Enzymol. 26*, 193 (1964).

THE GLYCOPEPTIDE LINKAGES OF EXTENSIN: O-D-GALACTOSYL SERINE AND O-L-ARABINOSYL HYDROXYPROLINE*

Derek T. A. Lamport

MSU/AEC Plant Research Laboratory
Michigan State University
East Lansing, Michigan 48823

Abstract

Primary cell walls contain extensin, a hydroxyproline-rich protein which appears to act as a crosslink between wall polysaccharides. The earlier demonstration of hydroxyproline arabinosides supported that hypothesis although we could not explain why there were only four sugar residues glycosidically attached to hydroxyproline. Recently we attempted to determine the amino acid sequence of extensin by taking advantage of the fact that treatment of cell walls at pH 1 for 1 hr. at 100°C completely cleaves the hydroxyproline-O-arabinose linkage and results in the previously protease-resistant protein becoming susceptible to tryptic attack. On separation of the tryptic peptides we observed that some of them possessed identical amino acid sequences yet were electrophoretically and

---

* This work was supported by AEC Contract AT911-1)-1338 and NSF Grant GB 24361.

and chromatographically distinct. Search for
non-ninhydrin reactive residues revealed D-galactose
as a component of most of the peptides and accounted
for the separations obtained. The glycopeptide
linkage was stable to $NaBH_4$ in 0.05 N NaOH at $4^{o}C$
but elimination of galactose occurred in 0.25 N NaOH
containing 1 M $NaBH_4$ at $50^{o}C$ and there was a
corresponding conversion of serine to alanine and
galactose to galactitol.

In view of these results we consider it most
likely that the major polysaccharide attachment
(probably arabinogalactan) to extensin is via the
serine hydroxyl group while the hydroxyproline
arabinosides could conceivably function to stablize
the hydroxyproline-rich extensin polypeptide backbone as a rigid rod.

Symbols

GLC = gas liquid chromatography

PTH = phenylthiohydantoin

Introduction

For many years there has been agreement that
the primary cell wall is a network. That view was
generally restricted to the cellulose microfibrils
(e.g. Roelofsen's multinet hypothesis (1) although
even in 1935 James Bonner (2) recognized the
probability of physical connections ("haftpunkte")
between the cellulose fibrils. Usually, however,
the non-cellulosic components of the wall have been
relegated to the position of matrix or "filler",
a view which hardly does justice to the undoubted
organization of these components.

The discovery of a novel hydroxyproline-rich
component in the cell wall led to the speculation
that wall polysaccharides might be cross linked

150

via this protein to form a network in which
covalent (?) linkages were cleaved in order to
allow cell extension (3, 4). Initial support for
this hypothesis came from the tenacity with which
the protein resisted attempts at removal from the
cell wall. In fact, nondegradative procedures
released very little of the wall protein. However,
enzymic degradation of the wall released
hydroxyproline-rich glycopeptides which also
contained arabinose and smaller amounts of galactose.
Analysis of these glycopeptides showed that the
arabinose was O-glycosidically linked to
hydroxyproline (5). This was further support for
the idea that wall protein could crosslink
polysaccharides, although precisely which
polysaccharides were involved was not known. In
addition, we assumed that the few galactose residues
also present in these glycopeptides were attached
to the arabinose residues. This was the simplest
view and it was based on the assumption that sugars
attached to serine would eliminate in the presence
of 0.05 N NaOH at 4°C overnight. That assumption
is not borne out by our recent results where we
successfully isolated hydroxyproline-rich
glycopeptides free from arabinose but containing
galactose.

Isolation of Galacto-glycopeptides

The hydroxyproline arabinosides isolated
previously are fairly stable to alkaline hydrolysis
but are completely hydrolysed at pH 1 after 1 hr
at 100°C. This is a convenient way to remove
arabinosyl residues from the hydroxyproline in the
wall protein without excessive peptide bond
cleavage. Thus, cell walls isolated from
tomato suspension cultures were refluxed at pH 1
for 1 hr. filtered and then treated with trypsin
at pH 8. Trypsin released about 45% of the hydroxyproline and this material was fractionated by gel
filtration on G-25 and by ion exchange chromato-

graphy on Aminex as previously described (6).
As a final purification step these tryptic peptides
were either electrophoresed at pH 1.9 or
chromatographed on Aminex A5.

Amino Acid Sequence of the Galacto-glycopeptides

The glycopeptides isolated were essentially resistant to further useful enzymic degradation, but
the presence of serine helped in sequence
determination (reported briefly previously) by
providing N-serine peptide bonds, which although
fairly stable to the pH 1 treatment were
hydrolysed by a short treatment with $6N$ HCl. Thus,
partial acid hydrolysis for 10 min in $6N$ HCl at
100°C showed that each glycopeptide contained at
least one unit of the pentapeptide Ser-Hyp$_4$. This
pentapeptide was identified unequivocally by the
isolation of the Ser-Hyp$_n$ homologous series from
n - 1 to n - 4. Sequencing via Edman degradation
and GLC of the PTH derivatives confirmed these
results, and also showed that peptides of identical
sequence (Fig. 1) were nevertheless electrophoretically and chromatographically distinct. This
microheterogeneity was traced to the presence
of a variable number of D-galactose residues in
each peptide.

Attachment of Galactose to Serine

In order to determine the site of galactose
attachment we selected glycopeptide S_2A_6 (Fig. 1,
sequence B) for close study. This peptide has two
galactose residues and two serine residues per mole.
This immediately suggested the possibility of an
O-galactosyl serine linkage. However, S_2A_6 did not
show the expected conversion of serine to alanine
when we attempted β-elimination in the presence
of NaBH$_4$ under the normal mild conditions
($0.5M$ NaOH + $0.2M$ NaBH$_4$ at 4°C for 18 hrs.) There-

A. Ser-Hyp-Hyp-Hyp-Hyp-Ser-Hyp-Ser-Hyp-Hyp-Hyp-Hyp-("Tyr"-Tyr)-Lys

   Exists with 3, 2, or 1 galactose residues.

B. Ser-Hyp-Hyp-Hyp-Hyp-Ser-Hyp-Lys

   Exists with 2, 1, or 0 galactose residues.

C. Ser-Hyp-Hyp-Hyp-Hyp-Thr-Hyp-Val-Tyr-Lys

   Exists with 1 or 0 galactose residues.

D. Ser-Hyp-Hyp-Hyp-Hyp-Lys

   Exists with 1 or 0 galactose residues.

E. Ser-Hyp-Hyp-Hyp-Hyp-Val-"Tyr"-Lys-Lys

   Exists with 1 or 0 galactose residues.

Fig. 1 Peptides isolated from acid-stripped tomato cell walls by treatment with trypsin.

___

fore we considered the possibility of galactose attachment to hydroxyproline, but hydrolysis of S_2A_6 in $Ba(OH)_2$ failed to provide evidence of galactosyl hydroxyproline*. Therefore we increased the severity of the β-elimination conditions and blacked the NH_2-terminal serine by maleylation of S_2A_6 in order to assist elimination of an O-substituent on the NH_2-terminus. Maleylated S_2A_6 at room temperature in the presence of NaOH and $NaBH_4$ showed significant conversion of serine to alanine after 24 hrs. and 48 hrs. (Table 1

___

* We have recently isolated galactosyl hydroxyproline after $Ba(OH)_2$ hydrolysis of Chlamydomonas cell wall(7).

expts. 2,3,4), while in experiment 2 (Table 1) there
was a good correlation between disappearance of
serine and loss of galactose from the peptide. Even
after 48 hrs at room temperature (Table 1 expt. 4)
only about 60% of the serine disappeared. However,
raising the temperature to 50° led to virtually
complete disappearance of serine within 5 hrs
(Fig. 2) while 45% of the theoretical yield of the
remaining alanine appeared in the reaction mixture.
Repurification of the eliminated peptide gave a
galactose-free peptide which contained 90% of the
theoretical yield of alanine. Essentially similar
results were obtained when S_2A_6 was succinylated
(Table 1, expt. 5) with the exception that lysine
yields were invariably decreased. When *unmodified*
peptide S_2A_6 (i.e. with free NH_2 terminus) was
treated with NaOH/NaBH$_4$ for 5 hrs at 50°C (Table 1,
expt. 1) slightly more than one residue of serine
disappeared, indicating that the peptide contained
two galactosyl serine residues one of which
eliminated more readily than the other. Treatment
of *unmodified* peptide S_2A_6 with 50 µl 0.25 N NaOH
1.0 M NaBH$_4$ under nitrogen for 5 hrs at 50°C followed
by acetylation and gas chromatography gave only one
product. This was identified as hexaacetyl galactitol. The yield of hexaacetyl galactitol was
53.5% of the total galactose in the peptide and in
conjunction with the results in Table 1 clearly
indicates the presence of a single galactose
residue O-glycosidically attached to each serine
residue.

   In an effort to obtain a better conversion
of serine than obtained by the NaOH/NaBH$_4$ technique,
we considered β-elimination in the presence of
sodium sulphite (8). Significant conversion of
serine to cysteic acid occurred when elimination
of succinylated peptide took place in the presence
of sodium sulphite at room temperature while an

PLANT CELL WALL POLYSACCHARIDES

Table 1

β-elimination of Ser-Hyp$_4$-Ser-Hyp-Lys (gal$_2$)

In the presence of NaBH$_4$

Expt.	Peptide Pretreatment	Elimination conditions	Composition [a]						
			Before Elimination			After Elimination			
			HYP	SER	LYS	HYP	SER	ALA	LYS
1.	none	0.2N NaOH 0.2N NaBH$_4$ 5 hr 50°	5	1.8	1.0	5	0.65	0.32	1.0
2.	maleylated[b]	0.2N NaOH 0.2M NaBH$_4$ 24 hr 25°	5	1.7	1.0	5	0.95	0.4	0.5
3.	maleylated	0.2N NaOH 1.0M NaBH$_4$ 24 hr 25°	5	1.7	1.0	5	1.2	0.07	1.0
4.	maleylated	0.2N NaOH 1.0M NaBH$_4$ 48 hr 25°	5	1.7	1.0	5	0.7	0.2	1.1
5.	succinylated	0.2N NaOH 0.2M NaBH$_4$ 5 hr 50°	5	1.5	0.5	5	0.09	0.55	0.52

[a] Expressed as amino acid residues relative to hydroxyproline. Figures are not corrected for losses during acid hydrolysis.

[b] In this preparation galactose before elimination was 2.0 residues and 1.3 residues after elimination.

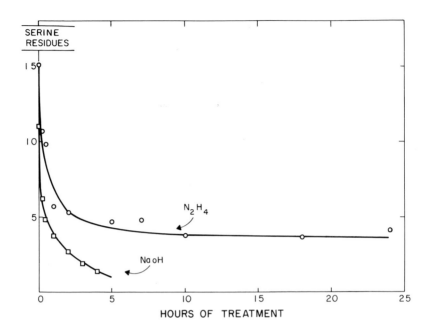

Fig. 2 Degradation of serine in cell walls after
treatment with NaOH or N_2H_4. Approximately 4 mg
samples of dry tomato cell wall (for hydrazinolysis)
or sycamore-maple cell wall (for NaOH treatment)
were heated at $110^{\circ}C$ for various times in a 1 ml
microvial containing either 400 μl $0.2N$ NaOH or
400 μl N_2H_4. Walls treated with NaOH were then
neutralised with 6 N HCl, dried by a stream of N_2
and then hydrolysed in $6N$ HCl, 18 hr at $110^{\circ}C$ for
subsequent amino acid analysis. Walls treated
with N_2H_4 were then heated at $50^{\circ}C$ and dried in a
stream of nitrogen followed by overnight vacuum
desiccation before hydrolysis in $6N$ HCl for 18 hr
at $110^{\circ}C$ for subsequent amino acid analysis.

The results are expressed as residues of
serine relative to 30 residues of hydroxyproline.
Note that NaOH treatment leads to a continual
decrease in serine while serine remains constant
after 10 hr of N_2H_4 treatment.

almost theoretical conversion occurred on increasing
the temperature to 50°C (Table 2, expt. 3).

Table 2

β-elimination of Ser-Hyp$_4$-Ser-Hyp-Lys (gal$_2$)

In the presence of Na$_2$SO$_3$

Peptide Pretreatment	Elimination conditions	Composition[a]						
		Before Elimination			After Elimination			
Expt.		HYP	SER	LYS	HYP	SER	CYST- EIC	LYS
1. none	0.25N NaOH 1.0M Na$_2$SO$_3$ 5 hr 50°	5	1.8	1.0	5	0.7	1.1	1.1
2. none	as in 1 but then cyanoethylated	5	1.8	1.0	5	0.0	0.8	0.0
3. succinylated	0.25N NaOH 1.0M Na$_2$SO$_3$ 5 hr 50°	5	1.5	0.5	5	0	1.6	0.4

[a] Expressed as amino acid residues relative to hydroxyproline. Figures are
not corrected for losses during acid hydrolysis.

Unmodified peptide (i.e. free NH$_2$ terminus)
treated with NaOH/Na$_2$SO$_3$ at 50°C gave 1.1 residues
of cysteic acid and 0.7 residues of serine
(Table 2 expt. 1). Further treatment of this
peptide with acrylonitrile at alkaline pH led to the
complete disappearance of serine (Table 2, expt. 2)
clearly showing that the internal serine residue
had undergone conversion to cysteic acid while the

157

NH$_2$-terminal serine-O-galactose residue was essentially stable under the elimination conditions used.

The most likely structure for glycopeptide S$_2$A$_6$ is therefore

$$
\begin{array}{cc}
\text{gal} & \text{gal} \\
| & | \\
\end{array}
$$
H$_2$N-Ser-Hyp-Hyp-Hyp-Hyp-Ser-Hyp-Lys-COOH which,

in view of the fact that most of the wall bound hydroxyproline is O-substituted by tetraarabinose, may have the following structure in the wall:

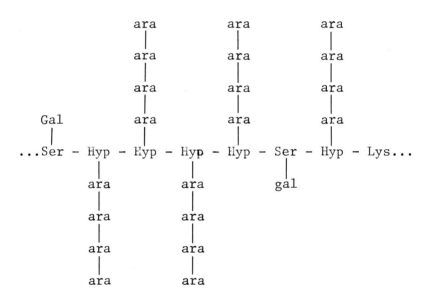

Discussion and Further Results

This structure raises a number of questions some of which are as follows:

How many serine residues are glycosylated in the wall?

Does this glycosylation represent
polysaccharide attached?

If so, which polysaccharide(s)?

If the major polysaccharide attachment is
via serine, do the hydroxyproline arabinosides
also represent sites of polysaccharide
attachment?

If not, what is the role of the hydroxyproline
arabinosides?

Which linkages in this glycoprotein might be
labilized to allow cell extension? Should
galactosyl-serine be considered as a candidate?

Currently we are attempting to answer a few
of these questions.

The degree of serine glycosylation seems
especially important. Unfortunately, however, in
the wall the stability of the galactosyl-serine
linkage to β-elimination seems even greater than in
the isolated peptides. This stability could
possibly arise from unfavorable peptide bond
conformation as well as polysaccharide attachment.
The linkage in the wall is in fact so stable that
alkaline conditions sufficiently vigorous to bring
about elimination (judged by serine loss from walls
treated with 0.2N NaOH at 110°C for various times,
Fig. 2) also eliminate the serine hydroxyl group in
non-glycosylated proteins such as trypsin and serum
albumin, although the rate of serine loss is somewhat lower in these proteins than in the wall.
However, this fact makes it impossible to measure
wall serine glycosylation via the usual reagent
(NaOH) for inducing elimination. Fortunately Heath
& Northcote (9) observed that wall protein could be
degraded by hydrazinolysis. They assumed that this

159

reagent did not cleave glycosidic linkages, at
least in their control material: cellulose. We
investigated the possibility that hydrazaine might
be sufficiently basic to induce elimination at an
elevated temperature. Hydrazinolysis of S_2A_6
followed by acid hydrolysis (to regenerate the
free amino acids from the amino acid hydrazides) led
to loss of nearly all the serine. Cell walls
treated with hydrazine also lost appreciable
amounts of serine (Fig. 2). Control experiments
with non-glycosylated proteins such as trypsin
or serum albumin to determine whether or not the
loss of serine was a nonspecific phenomenon similar
to the loss observed in the presence of NaOH at
an elevated temperature showed that *serine was
quite stable to hydrazinolysis* both in the presence
and absence of carbohydrate (Tables 3 and 4).
Clearly, therefore, the molecular environment
of serine residues in the wall renders them
susceptible to degradation in the presence of
hydrazine. Because the glycosylated serine
residues of peptide S_2A_6 are unstable to
hydrazinolysis we infer that the wall serine
residues which are lost during hydrazinolysis
are also glycosylated. Thus, we do have a crude
method for measuring serine glycosylation in the wall
and we can therefore attempt to answer the question
about its possible role as a labile linkage during
growth. At the present time preliminary experiments
indicate that the number of glycosylated serine
residues might be a function of growth. We
prepared cell walls from tomato suspension
cultures of various ages and measured the ratio of
hydrazine-stable serine to hydroxyproline
(Fig. 3).

Table 3

		Serum Albumin	Serum Albumin & Carbohydrate	% Change in Presence of Carbohydrate	Serum Albumin & Carbohydrate in N_2H_4	% Change in Presence of N_2H_4 & Carbohydrate
		A nmoles	B nmoles	C	D nmoles	E
	ASP	117	84	-28	85	-27
	THR	69	47	-32	39	-43
	SER	52	44	-15	45	-13
	GLU	163	119	-27	142	-13
	PRO	nd.	nd.	-	nd.	-
	GLY	35	26	-26	53	+51
	ALA	94	69	-27	86	- 9
	VAL	73	52	-29	64	-12
1/2	CYST	68	37	-46	9	-32
	MET	7	5	-29	4	-43
	ILU	26	18	-31	20	-23
	LEU	129	94	-27	114	-11
	TYR	41	15	-63	22	-46
	PHE	57	40	-30	51	-11
	ORN	0	0	-	31	-
	LYS	118	90	-24	103	-13
	HIS	35	25	-29	24	-31
	ARG	45	28	-38	0	-100

Equal amounts of serum albumin were hydrolysed in 6N HCl (18 hr at 110°) before (**B**) and after hydrazinolysis (D) (10 hrs in N_2H_4 at 110°) in the presence of a mixture of sugars similar to that present in the tomato cell wall. A control hydrolysis (A) in the absence of carbohydrate indicates the extent of degradation for each amino acid due to the presence of carbohydrate during acid hydrolysis. Sugar molar ratios used were: arabinose 4.2; xylose 1.8; galactose 4.6; glucose 10.0; mannose 2.2; galactoronic acid 2.3. When the protein was hydrolysed in the presence of carbohydrate the mixture consisted of 20% (by weight) serum albumin, 80% sugars. The serum albumin was a crystalline preparation obtained from the Sigma Chemical Company.

161

Table 4

	Trypsin	Trypsin & Carbohydrate	% Change in Presence of Carbohydrate	Trypsin & Carbohydrate & N_2H_4	% Change in Presence of N_2H_4 & Carbohydrate
	A nmoles	B nmoles	C	D nmoles	D
ASP	129	99	−23	92	−29
THR	53	42	−21	36	−32
SER	180	143	−21	149	−17
GLU	85	68	−20	71	−16
PRO	28	25	−11	37	+32
GLY	152	117	−23	148	− 3
ALA	90	76	−16	72	−20
VAL	79	61	−14	72	− 9
1/2 CYST	68	39	−43	11	−84
MET	11	8	−27	7	−36
ILU	75	60	−20	52	−31
LEU	87	67	−23	73	−16
TYR	59	24	−59	30	−49
PHE	21	16	−24	18	−14
ORN	0	0	0	8	+
LYS	78	61	−22	67	−14
HIS	16	11	−31	10	−38
ARG	11	7	−36	0	−100

Trypsin was hydrolysed in 6N HCl under conditions similar to those described for serum albumin in Table 3. The trypsin was a crystalline preparation (TRTPCK) obtained from the Worthington Biochemical Corporation.

162

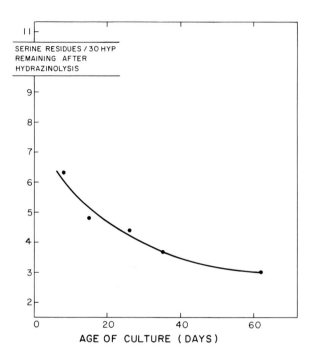

Fig. 3 Hydrazine-labile serine residues in cell walls isolated at various stages of growth. Cell walls were isolated from tomato suspension cultures at various stages of growth and the serine-hydroxy-proline ratio was measured after hydrazinolysis as described in Fig. 2.

The hydrazine-stable serine *decreased* with culture age and one interpretation is that this indicates an *increase* in glycosylated serine. This would be consistent with the general extinsin hypothesis which calls for increased cross linking as cell extension decreases.

163

Acknowledgements

I thank Laura Katona and Sandra Roerig for their highly skilled assistance in this work.

References

1. P.A. Roelofsen and A.L. Houwink, *Acta Bot. Neerl.* 2, 218, (1953).
2. J. Bonner *Jahrb. wiss. Botan.* 82, 377 (1935).
3. D.T.A. Lamport, *J. Biol. Chem.* 238, 1438 (1963).
4. D.T.A. Lamport, *Advan. Bot. Res.* 2, 151 (1965).
5. D.T.A. Lamport, *Biochemistry* 8, 1155 (1969).
6. D.T.A. Lamport, *30th Symp. Soc. Develop. Biol.* in press (1972).
7. D.H. Miller, D.T.A. Lamport and M. Miller, *Science.* 176, 918 (1972).
8. S. Harbon, G. Herman and H. Clauser, *European J. Biochem.* 4, 265 (1972).
9. M. F. Heath and D. H. Northcote, *Biochem. J.* 125, 953 (1971).

THE BIOSYNTHESIS AND SECRETION OF HYDROXYPROLINE-RICH CELL WALL GLYCOPROTEINS*

D. Sadava and Maarten J. Chrispeels**

Department of Biology, John Muir College
University of California at San Diego
La Jolla, California   92037

Abstract

Carrot-root phloem parenchyma cells synthesize
and secrete to the cell wall a hydroxyproline-rich
glycoprotein. The entire process takes about fifteen minutes. During the first three minutes, the
polypeptide backbone is synthesized on polyribosomes.
After this time, certain proline residues are
hydroxylated via a soluble peptidyl proline hydroxylase. The modified protein then becomes associated
with a cytoplasmic organelle and is glycosylated. A
particulate UDP arabinosyl transferase catalyzes
arabinose addition to hydroxyproline residues. Other
glycosylations may occur at this stage. During the
next 7 to 10 minutes, the completed glycoprotein is
secreted to the cell wall. Secretion is an energy-
requiring process which is mediated by membranous

---

*This work was supported primarily by the U.S.
Atomic Energy Commission [Contract AT (04-3)-34,
Project 159] and by the U.S. National Science Found-
ation. (GB 302235).

**To whom inquiries should be addressed.

organelles. The organelles which contain the UDP arabinosyl transferase are probably identical with those which transport the glycoprotein to the cell membrane.

Introduction

The structure of the hydroxyproline-rich glycoprotein of the primary cell wall is described elsewhere in this volume (Lamport). In this contribution, we will review information concerning the cytoplasmic biosynthesis of this glycoprotein and its secretion across the cell membrane to the cell wall. Several aspects of this work are discussed in greater detail in another review (5).

Most of our experiments have been performed on thin disks of phloem paranchyma from the storage roots of carrots (*Daucus carota* cv.). When such disks are incubated in a moist environment, they undergo a number of physiological changes collectively termed "aging". Increases in respiration (31), protein synthesis (11) and net cell organelles (13) have been noted. In the aging conditions we use (2), there is no division, expansion, or differentiation of the carrot cells. However, there is a considerable increase in cell wall-bound hydroxyproline (2). Thus this system of aging carrot disks provides a homogeneous cell population active in the biosynthesis and secretion of cell wall hydroxyproline glycoproteins.

The specific presence of hydroxyproline in the cell wall glycoprotein has provided a convenient tag with which to follow its assembly and secretion. Experiments using radioactive labels have led to a view that, following the synthesis of the protein backbone, two sequential post-translational modifications of the protein occur. First, certain proline residues are hydroxylated. Then, many of the newly-formed

hydroxyproline residues are glycosylated with arabinose. The completed glycoprotein is then secreted
across the cell membrane into the cell wall matrix.

Hydroxylation of Peptidyl Proline

The absence of a free hydroxyproline pool in
plant cells and the fact that exogenous hydroxyproline
was not incorporated into proteins led Steward to
suggest that hydroxyproline arises from peptidyl proline (24). Subsequent investigations have confirmed
this hypothesis.

There is a lag time between the appearance of
exogenous radioactive proline in proteins and the
appearance therein of radioactive hydroxyproline (7,
23, 24). In carrot disks, this lag is only a few
minutes (2). However, this is sufficient time for the
incorporation of the radioactive proline residues and
completion of the polypeptide chain, since published
times for the completion a protein in eukaryotes
are of the order of one minute (26, 34). If hydroxyproline formation begins at or after polypeptide
chain completion, one would not expect hydroxyproline
in nascent polypeptides still attached to the
polyribosome. Indeed, when carrot disk or tobacco
tissues are exposed for short times (7 minutes)
to radioactive proline and the nascent polypeptides
attached to their polyribosomes isolated, no radioactive
hydroxyproline is found (27). By this time, the
cytoplasmic protein show 10% of their ratioactive
proline residues as hydroxyproline. Thus the initial
modification of the future cell wall glycoprotein,
that of proline hydroxylation, occurs on completed
polypeptides either in the soluble cytoplasm or
on the outside surface of membranous organelles.

Considerable *in vivo* evidence indicates the nature of peptidyl hydroxyproline synthesis. The process has a temperature coefficient of 2.2, indicating
an enzymatic reaction (3). Molecular oxygen is the

167

source of the hydroxyl oxygen, as is shown from $^{18}O_2$ mass spectrometry (18, 30). Peptidyl proline residues tritiated in the trans-3 and trans-4 positions lose the 4-trans tritium upon their conversion to hydroxyproline (19). Iron is probably involved in the reaction, since the chelator α,α'-dipyridyl blocks hydroxyproline synthesis (14), and this inhibition can be overcome by exogenous ferrous ions (3).

A cell-free system which has the above characteristics has been described for carrot disks (28). The enzyme, peptidyl proline hydroxylase, will catalize the *in vitro* hydroxylation of peptidyl and not free proline. It requires Fe^{++}, ascorbate, α--ketoglutarate and O_2 for activity. During the reaction, the 4-trans proton of proline is displaced by the incoming hydroxyl group. The plant enzyme has many characteristics in common with collagen proline hydroxylase from animal cells (25). Indeed, it will catalyze the hydroxylation of proline residues in protocollagen. The plant enzyme activity is localized in the soluble cytoplasm. As yet, it is relatively impure and therefore information on its mechanism is unavailable. In terms of glycoprotein synthesis, it would be interesting to know why, and indeed which, proline residues are selected for hydroxylation.

Glycosylation of Peptidyl Hydroxyproline

Following proline hydroxylation, a second series of post-translational modifications occurs on the protein, that of glycosylation. Kinetic experiments using radioactive proline indicate that in carrot disks glycosylation begins about four minutes after proline hydroxylation (4). The major reaction is the glycosylation of the newly-made hydroxyprolines with arabinose. Hydroxyproline-arabinosides have been isolated after alkaline hydrolysis of cytoplasmic and cell wall proteins of many plants (1, 21). The initial point of attachment of an O-glycosidic bond from

168

the 4-hydroxyl of the imino acid to the 1-carbon of
the pentose sugar (20). Up to three more arabinose
residues may be attached to the initial one.

Recently, an enzyme system catalyzing the *in
vitro* glycosylation of peptidyl hydroxyproline was
described. (17). This system, isolated from cultured sycamore cells, is particulate and sediments
when the crude extract is centrifuged at 37,000g. It
catalyzes the transfer of sugar from UDP-arabinose
to hydroxyproline in an endogenous acceptor present
in the particle. The hydroxyproline-arabinosides thus
formed are indistinguishable from those present in the
cell wall. As yet, this activity also is impure; it
can also catalyze the incorporation of arabinose into
larger polysaccharides. Nevertheless, its particulate
nature suggests that the hydroxylated protein must
become associated with a membranous compartment in the
cell to be glycosylated. In animal cells, the Golgi
apparatus contains many of the glycosidases which
modify secreted proteins (22, 23); this may also be
the case in plants.

Other glycosylation reactions may occur before
the glycoprotein is completed and ready to be secreted. Glycopeptides isolated from digestion of
cell walls contain galactose residues glycosidically
linked to serine (Lamport, this volume). Whether this
linkage is made in the cytoplasm or the cell wall remains to be determined.

We have isolated from the cytoplasmic organelles
of carrot disks an apparently completed cell wall
glycoprotein prior to its secretion. A glycoprotein
with similar properties can be extracted from the
cell wall (1). This component accounts for one-third
of the hydroxyproline that these cells secrete to the
wall. The glycoprotein is of high molecular weight
(about 200,000 as determined by gel filtration) and
is about 60% protein and 40% carbohydrate. The protein moiety is rich in hydroxyproline (10 mole percent)
and serine (19 mole percent). The carbohydrate

moiety contains arabinose linked to hydroxyproline. However, this cell wall protein does not contain measurable galactose, suggesting that the serine-O-galactose linkage (Lamport, this volume) may be formed within the cell wall.

Secretion of the Glycoprotein

After its glycosylation occurs at the particulate site, the glycoprotein is secreted through the cell membrane into the cell wall matrix. Kinetic experiments on carrot disks indicate that the glycoprotein arrives in the wall about seven minutes after glycosylation. This would mean that the total time for synthesis and secretion of the carrot cell wall glycoprotein is about fifteen minutes (4).

Secretion is an independent event and does not depend upon the continuation of the events preceeding it. For example, pulse-chase experiments indicate that concomitant protein synthesis is not necessary for secretion. Cycloheximide does not inhibit the transport to the wall of those glycoproteins made before its addition (10). Moreover, proline hydroxylation (and hence the bulk of glycosylation with arabinose) is not a prerequisite for secretion. Kinetic studies show that the synthesis and secretion of cell wall proteins continues in the presence of α,α'-dipyridyl. This chelator blocks proline hydroxylation without affecting overall protein synthesis. This situation contrasts with its analogue in animal cells. If collagen-synthesizing cells are challenged with dipyridyl (blocking lysine hydroxylation and glycosylation), collagen secretion stops and the unhydroxylated proteins accumulate within the cell (35). Indeed, in animal cells protein glycosylation and protein secretion are intimately related (12). In the case of the plant glycoprotein, the indirect effect of dipyridyl on glycosylation may not be complete: serine-O-galactose may be present, for example, thus permitting the secretion of a glycosylated macromolecule.

170

Glycoprotein secretion in plants does require metabolic energy. Uncouplers of oxidative phosphorylation and electron transport inhibit the secretion in carrot disks (10). Similar observations have been made on other plant (32) and animal (16) cells.

Membranous organelles are involved in the secretion of the hydroxyproline-rich glycoprotein, as they are in glycoprotein secretion and in animal cells (15, for example). That the glycoprotein is initially in a membranous compartment is indicated by the particulate nature of the glycosylation reactions. Identification of the organelles participating in secretion involves their separation by gradient centrifugation and determination of the organelle(s) with which the glycoprotein is transiently associated. Data are available on this subject from cultured sycamore cells (9) and from research in progress on carrot disks (M. G. Gardiner, our laboratory). Both studies are inconclusive as to the nature of the organelles involved in secretion. The sycamore study used velocity sedimentation and discontinuous gradients. The results show three separate cell fractions with which radioactively labelled glycoprotein is transiently associated. The one component common to the three fractions is a smooth membranous organelle. The carrot study uses isopycnic and rate zonal sedimentations to separate organelles. Its results show a single fraction containing, transiently, both hydroxyproline and arabinose. This fraction contains most of the cellular UDP arabinosyl transferase activity, suggesting that it is involved in glycosylation. Its density (1.15 g/cc.) and enzyme properties (inosine diphosphatase) are characteristic of Golgi bodies (22). However, the positive identification of the organelles involved in, and the precise mode of glycoprotein secretion remain challenging problems.

Regulation of Biosynthesis and Secretion

During aging of carrot disks (2) and the ces-
sation of elongation in pea epicotyls (8), there is
a substantial accumulation of cell wall glycoproteins
relative to other cell proteins. Since the hydroxyproline-rich component is stable (24), it is unlikely that the cell regulates the amount of it by
differential breakdown. Rather, the regulation of
cell wall protein acculumation must be within the
sequential scheme discussed here.

Studies of both carrot disks (5) and pea epicotyls
(29) indicate that the regulation of the amount of the
cell wall glycoprotein lies at the level of protein
synthesis. The hydroxylating, glycosylating and secretory systems do not appear to be limiting the ac-
culation of glycoproteins in the wall. Rather, the
amount of glycoproteins made increases with respect
to other proteins. In aging carrot disks, the ratio
of bound radioactive proline residues as hydroxyproline rises from less than 1% in freshly cut disks
to 25% after 20 hr aging. As pea epicotyls cease
active elongation, this index of hydroxyproline synthesis rises from 19% to 33%. It appears that the
pool of unhydroxylated proteins destined to become
cell wall glycoprotein is augmented to increase
the glycoprotein level in the wall.

Conclusion

We have presented a broad scheme for the sequential synthesis and secretion of the hydroxyline-rich
cell wall glycoprotein. However, many questions as
to the details of this scheme remain to be answered.
Is the protein preferentially synthesized on rough
endoplasmic reticulum? What determines which prolines
are hydroxylated or hydroxyprolines glycosylated?
Which organelles are involved in glycosylation and
secretion? How does the glycoprotein get across the
cell membrane? How is it built into the wall? What

is its function in the wall? These problems will no doubt occupy our and others' laboratories in the near future.

References

1. M. Brysk and M. J. Chrispeels, *Biochim. Biophys. Acta* 257, 421 (1972).
2. M. J. Chrispeels, *Plant Physiol.* 44, 1187 (1969).
3. M. J. Chrispeels, *Plant Physiol.* 45, 223 (1970).
4. M. J. Chrispeels, *Biochem. Biophys. Res. Commun.* 39, 732 (1970).
5. M. J. Chrispeels, *Develop. Biol. Suppl.* 5, (in press 1972).
6. M. J. Chrispeels, *Plant Physiol.* (submitted).
7. R. Cleland, *Plant Physiol.* 43, 865 (1968).
8. R. Cleland and A. Karlsnes, *Plant Physiol.* 42, 669 (1967).
9. W. Dashek, *Plant Physiol.* 46, 831 (1970).
10. M. Doerschug and M. J. Chrispeels, *Plant Physiol.* 46, 363 (1970).
11. R. Ellis and I. R. MacDonald, *Plant Physiol.* 42, 1297 (1967).
12. E. H. Eylar, *J. Theoret. Biol.* 10, 89 (1965).
13. L. Fowke and G. Setterfield in *Biochemistry and Physiology of Plant Growth Substances*, (F. Wightman and G. Setterfield, eds.) Runge Press, Ottawa (1968) p. 581.
14. J. Holleman, *Proc. Natl. Acad. Sci. U.S.* 57, 50 (1967).
15. J. Jamieson and G. Palade, *J. Cell Biol.* 34, 577 (1967).
16. J. Jamieson and G. Palade, *J. Cell Biol.* 39, 589 (1968).
17. A. Karr, *Plant Physiol.* 50, 275 (1972).
18. D. T. A. Lamport, *J. Biol. Chem.* 238, 1438 (1963).
19. D. T. A. Lamport, *Nature* 202, 293 (1964).
20. D. T. A. Lamport, *Nature* 216, 1322 (1967).
21. D. T. A. Lamport and D. Miller, *Plant Physiol.* 48, 454 (1971).
22. D. J. Morré, L. Merlin and T. Keenan, *Biochem. Biophys. Res. Commun.* 37, 813 (1969).

23. A. C. Olson, *Plant Physiol.* *39*, 543 (1964).
24. J. K. Pollard and F. C. Steward, *J. Exp. Bot. 10*, 17 (1959).
25. J. Rosenbloom and D. J. Prockop in *Repair and Regeneration*, (Dunphy and Van Winkle, eds.) McGraw-Hill, New York (1968) p. 117.
26. J. Rosenbloom, R. Bhatnagar and D. J. Prockop, *Biochim. Biophys. Acta 149*, 259 (1967).
27. D. Sadava and M. J. Chrispeels, *Biochemistry 10*, 4290 (1970).
28. D. Sadava and M. J. Chrispeels, *Biochim. Biophys. Acta 227*, 278 (1971).
29. D. Sadava, F. Walker and M. J. Chrispeels, *Dev. Biol.* in press (1972).
30. E. R. Stout and G. J. Fritz, *Plant Physiol. 41*, 197 (1966).
31. K. V. Thimann, C. S. Yocum and D. P. Hackett, *Arch. Biochem. Biophys.* 53, 240 (1954).
32. J. E. Varner and R. Mense, *Plant Physiol. 49*, 187 (1972).
33. R. Wagner and M. Cynkin, *J. Biol. Chem. 246*, 143 (1971).
34. R. Winslow and V. I. Ingram, *J. Biol. Chem. 241*, 1144 (1966).
35. K. Juva, D. J. Prockop, G. W. Cooper and J. W. Lash, *Science 152*, 92 (1966).

PISTIL SECRETION PRODUCT AND POLLEN
TUBE WALL FORMATION*

F. Loewus and C. Labarca**

Department of Biology
State University of New York at Buffalo
Buffalo, New York   14214

Abstract

The glandular tissue lining the stigma surface
and stylar canal of lily pistils secretes an exudate in which about 90 to 95% of the solute is carbohydrate, predominately polysaccharide. This secretion facilitates pollen tube development following
pollination and has long been regarded as a potential source of tube wall carbohydrate. Experimental
evidence for this nutritional dependence has now
been obtained. When detached lily pistils are
pulsed with D-glucose-^{14}C, about 5% of the label is
converted to stigmatic exudate. Pollen tubes recovered from pistils directly labeled with D-glucose-
^{14}C obtain as much as 70% of their carbohydrate from
labeled exudate. The process involves actual transfer of polysaccharide fragments from the stylar canal

*This investigation was supported by NIH grant
number GM-12422 from the National Institute of
General Medical Sciences.
**Present address: Facultad de Ciencias, Univer-
sidad de Chile, Casilla 653, Santiago, Chile.

into pollen tube cytoplasm and eventual deposition
of newly formed tube wall that consists of polysaccharide reconstructed from monosaccharide residues
and larger fragments of the exudate polysaccharide.
The arabinogalactan portion of exudate appears to
undergo the most extensive breakdown to furnish
hexose units for biosynthesis of pollen tube glucans.
Results suggest that cell wall biosynthesis, utilizing polysaccharide fragments derived from secreting
tissues and cells, may play a significant role in
higher plant cell wall biogenesis.

Introduction

During initial stages of pollen tube formation,
stored reserves present in the pollen grain provide
material for tube wall biosynthesis. As pollen
tubes penetrate the pistil, a second nutritional
source of wall material is gradually exposed (1).
In solid pistils, pollen tubes penetrate a cytologi-
cally distinct region of the style referred to as
"transmitting tissue", a tissue in which cells are
embedded in a dense carbohydrate matrix. Dr. M. Kroh
has initiated studies on the composition of intercellular substance from pistil transmitting tissue
of *Petunia* and reports her results in the following
chapter.

In pistils containing hollow styles, pollen
tubes grow superficially, that is, on the surface of
specialized cells which cover the stigma and line
the stylar canal. An exudate secreted by these
cells surrounds the pollen tubes. The precise nature
of pollen tube metabolism as it relates to utilization
of exudate in hollow styles has received little attention yet any attempt to fully understand the biochemical basis of genetic barriers to fertilization
must include these nutritional considerations.

Composition of Pistil Secretion Product

We have been using the floral system of *Lilium longiflorum*, the Easter lily, to study the nutritional relationship between pistil secretion product and pollen tube wall formation. A photograph of the flower, with that portion of tepals and stamens on the side of the flower facing the viewer removed, is seen in the upper half of Fig. 1. The picture, taken about 3 days after anthesis or bud opening, shows a drop of stigmatic exudate clinging to a lobe of the stigma. It is this secretion product that provides the nutritional link between pistil and growing pollen tube.

When stigmatic exudate is separated on a Sephadex G-100 gel column, most of the carbohydrate, about 90% of the dry weight of the exudate, elutes with 0.01 M acetic acid as a very high molecular weight fraction, G-100-I, partially excluded from the gel. A low molecular weight fraction, G-100-II, rich in oligosaccharides, accounts for less than 5% of the exudate. A typical profile of carbohydrate. in fractions eluted from Sephadex G-100 is given in Fig 2.

G-100-I is a mixture of acidic polysaccharides and/or glycoprotein which upon acid hydrolysis yields the monosaccharide composition given in Table 1. Glucuronic acid exceeds galacturonic acid over 3-fold. This fraction is also rich in galactose, arabinose and rhamnose. In these respects it resembles gum exudates of plant origin more closely than pectic substance.

When detached pistils are placed in solutions containing D-glucose-1-^{14}C, D-glucose-U-^{14}C, or *myo*-inositol-U-^{14}C, about 5% of the label (Table 2) is converted to labeled G-100-I and a lesser amount to G-100-II. For reasons not understood as yet,

177

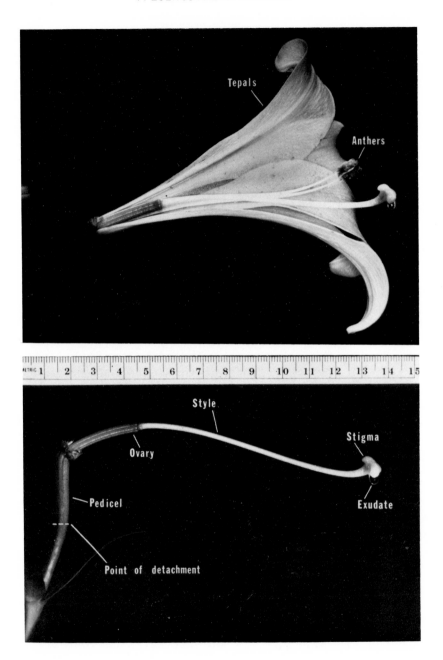

Fig. 1

Fig. 1 Two views of a flower of *Lilium longi-florum*, cv. Ace. Upper photograph: side view with portions of tepals and 4 stamens removed. Lower photograph: same side view with all tepals and stamens removed. Note the accummulation of exudate at the stigma. The ruler is numbered in cm.

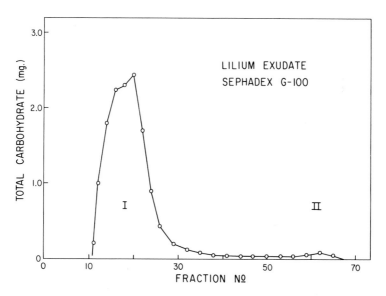

Fig. 2 Gel filtration profile of stigmatic exudate. Analysis of carbohydrate was made with phenol-sulfuric acid reagent.

TABLE 1

Composition of G-100 Fraction

Component	Amount
	%
Glucuronic acid	9.5
Galacturonic acid	2.5
Galactose	30
Arabinose	28
Rhamnose	12.5
Total carbohydrate	82.5
Protein	6.6

TABLE 2

Conversion of Labeled Carbohydrate or Amino Acid

to Stigmatic Exudate in Detached Lily Pistils

Source of label	Radioactivity supplied per pistil	Radioactivity recovered in stigmatic exudate
	μC	%
myo-Inositol-2-^{3}H	27	1.0
myo-Inositol-U-^{14}C	10	5.5
	12.5	5.9
	12.5	8.0
	12.5	3.9
D-Glucose-1-^{14}C	18	5.0
L-Proline-U-^{14}C	10	1.1

myo-inositol-2-^{3}H and L-proline-U-^{14}C are less effective sources of label to exudate. When the G-100-I fractions of these labeled exudates are hydrolyzed with acid and chromatographed on paper, results listed in Table 3 are observed. When labeled myo-

TABLE 3

Distribution of Radioactivity among Components of
Acid-hydrolyzed G-100-I Fraction of Stigmatic Exudate

Component*	myo-Inositol 2-^{3}H	myo-Inositol U-^{14}C	D-Glucose 1-^{14}C	L-Proline U-^{14}C
	%	%	%	%
Uronic acid (origin)	15.7	19.0	11.8	17.2
Mannose	1.9	<1	1.0	1.5
Galactose	4.9	16.2	55.6	53.5
Arabinose	68.7	53.2	17.3	21.3
Xylose (fucose)	2.6	<1	<1	1.5
Rhamnose	1.0	3.1	5.1	4.9
Unknown (eluted)	5.2	8.9	8.5	1

* Position on paper after chromatography in ethyl acetate-
-pyridine-water (8:2:1, v/v).

inositol is supplied, the bulk of the label is recovered in uronic acids and arabinose. With labeled glucose, the bulk of the radioactivity appears
in galactose but all saccharide residues are labeled. Paper chromatography reveals no labeled glucose
residues in G-100-I but gas-liquid chromatography of
the alditol acetates of saccharides after acid
hydrolysis of G-100-I reveals the presence of a trace
of glucose (2, 3).

To summarize, detached lily pistils produce
an exudate containing about 10% solids, largely
polysaccharide in composition. The exudate has a
gross saccharide composition resembling gum exudate
of plant origin.

The Detached Lily Pistil as an Experimental Tool

As a model for the study of cell wall formation,
the pollinated pistil has certain practical advantages. Pollen tube growth is rapid, easily controlled and experimentally manipulative. A lily pistil,
stripped of tepals and stamens as shown in the
lower half of Fig. 1, is detached at the pedicel
and transferred to a small vial. The detached
pistil will survive for 6 to 8 days under the proper
environmental conditions of light, temperature and
humidity. Up to 10 mg of pollen is easily spread
over the stigma surface. Germination is rapid.
Pollen tubes will grow from stigma to ovary, a distance of about 10 cm, in 72 hr at 27 C. By using
batches of pistils, sufficient pollen tube tissue
for chemical as well as isotopic studies can be
dissected from the pistils.

When a detached pistil is placed in a vial containing a labeled carbohydrate such as myo-inositol-
$2-^3H$, label is rapidly transferred through the vascular system into the upper region of the pistil,
primarily the stigma and tissues which surround the
stylar canal (4). Some of the label is converted to

exudate. An autoradiograph of a cross-section through the style of such a pistil about 2 days after pollination is seen in Fig. 3. Darkened regions in the photograph correspond to tritium labeled polysaccharide produced by metabolism of *myo*-inositol oxidation pathway (see Loewus, Chen and Loewus, this volume). Vascular tissue, located in the apices of the crudely triangular cross-section, contains much tritium labeled polysaccharide in the phloem but very little in the xylem. Labeled exudate fills the canal. Labeled pollen tubes, seen in cross-section in the canal, are embedded in the exudate. Cells bordering the canal, referred to in the earlier literature as stigmatoid cells but now called canal cells (5), are regarded as a special class of secreting cells, the transfer cells (6). Canal cells polarize the movement of carbohydrate reserves from vascular tissues, converting these reserves into exudate and transferring the exudate into the stylar canal through a highly specialized, secreting cell wall. The latter is easily seen in the autoradiograph as a thick dark scalloped border ringing the canal.

The problem of pollen tube wall biogenesis, particularly its relationship to secretion products of the pistil can be studied in many ways as outlined in the diagram in Fig. 4. Preliminary studies involved tube wall formation in media containing stigmatic exudate or exudate fractions (5, see also Fig. 6 in Loewus, Chen and Loewus, this volume). To obtain a more quantitative view, studies were undertaken involving the growth of pollen tubes on pistils bearing labeled, injected exudate (3). Our most recent work involves pollen tubes recovered from detached pistils which were labeled directly (Labaraca and Loewus, submitted for publication).

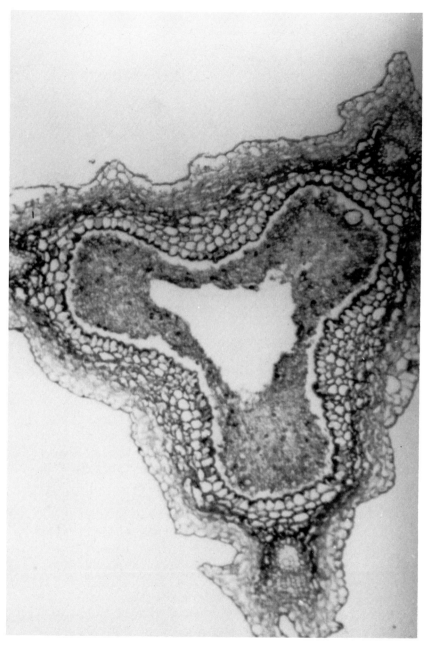

Fig. 3 Autoradiograph of a cross-section of the style from a *myo*-inositol 2-³H labeled pistil of *L. longiflorum* at approximately 100 magnifications.

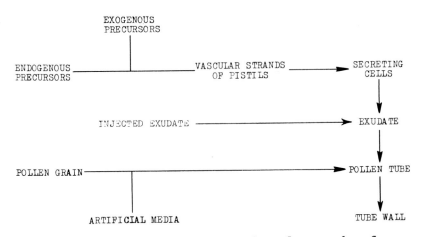

Fig. 4 Experimental approaches for study of pollen tube wall formation.

Pollen Tube Wall Formation in Pistils
Bearing Injected Labeled Exudate

The G-100-I fraction of exudate recovered from detached pistils labeled with D-glucose-1-^{14}C was injected into freshly pollinated pistils. Three days after injection, pollen tubes were recovered from pistils as shown in Fig. 5. First the style was detached from the ovary. Then the top surface of the stigma was sliced so as to sever pollen tubes from the grains which remained embedded in the stigma surface. The style section was submerged in 0.3 M pentaerythritol, split longitudinally, and pollen tubes teased out of the canal with a dissecting needle. Exudate adhering to the tubes was removed by further rinsing with fresh pentaerythritol solution. Finally, the tubes were ground in water to yield an insoluble wall fraction and a turbid supernatant which we have called the soluble fraction. Together, these fractions accounted for 6.1% of the injected radioactivity. The wall fraction had a specific radioactivity roughly 25% of that found in

DISTRIBUTION OF LABEL* IN PISTIL AND POLLEN 83 HOURS AFTER INJECTION

	Disecting medium	Pollen tubes	Style	Stigma and pollen grains	Total
Wash (exudate)	15.9	6.3	6.0	29.7	57.9
Homogenate:					
soluble	-	2.4	2.7	5.3	10.4
insoluble	-	3.7	1.5	5.7	10.9
					79.2

* As % of label injected into the pistils.

Fig. 5 Dissection of pollinated pistil for recovery of labeled pollen tubes. Data is taken from a cv. Ace pistil injected with G-100-I fraction of stigmatic exudate from D-glucose-1-^{14}C labeled cv. Ace pistils. The injected pistil was cross-pollinated with 10 mg of cv. No. 44 pollen. Only the stigma-style portion of the pistil was analyzed (3).

---

exudate adhering to the excised pollen tubes. In other words, at least 25% of the carbohydrate in the tube wall was derived from exudate.

During initial fractionation of pollen tubes, about 40% of the radioactivity remained in the turbid supernatant, the soluble fraction. Further centrifugation at 20,000g produced a pellet containing one-fourth of the label and a clear supernatant accounting for the remaining radioactivity. The latter was fractionated on Sephadex G-100 into high, intermediate and low molecular weight fractions with the ^{14}C distribution, 6, 22 and 2% respectively, based on the total radioactivity in the pollen tubes before grinding.

185

Acid hydrolysis of each fraction, as well as the pellet and the wall, followed by paper chromatography
revealed the distribution of label among saccharide
components shown in Table 4. Note that although

TABLE 4

Distribution of Radioactivity among Saccharide Components of Pollen

Cytoplasm and Wall Fractions from Pistils Injected with Labeled

Stigmatic Exudate (D-Glucose-1-^{14}C derived).

Component*	G-100-I Fraction (Injected exudate)	Cytoplasmic Fractions				Tube Wall
		High MW	Intermed MW	Low MW	Pellet	
	%	%	%	%	%	%
Origin	12	35	8	40	24	27
Galactose	58	30	63	12	30	10
Glucose	-	1	1	25	16	19
Mannose	1	8	1	2	4	2
Arabinose	15	10	20	6	12	30
Xylose-fucose	1	7	1	2	2	3
Rhamnose	5	1	3	4	6	5
Unknown (eluted)	8	7	5	9	6	3

* Position on paper after chromatography in ethyl acetate-pyridine-water (8:2:1, v/v).

D-glucose-1-^{14}C was used to label pistils which
produced labeled exudate, the exudate was virtually
devoid of labeled glucose. Moveover, very little
labeled glucose was detected in either the high or
intermediate molecular weight fractions of pollen
tube cytoplasm. Significant amounts of labeled
glucose was detected in the low molecular weight
fraction of the cytoplasm, the pellet and the tube
wall. Other results, not presented here, indicated
that this newly formed glucose was derived from
galactose residues, especially those in the intermediate fraction from tube cytoplasm (3). In

186

contrast to the G-100-I fraction in which 75% of the uronic acid was glucuronic acid, all of the uronic acid found in pollen tube walls was galacturonic acid.

Our studies suggest that at least two kinds of polysaccharide are present in droplets of exudate recovered from the stigma, one a very high molecular weight component containing traces of xylose and mannose and the other, a polymolecular mixture of intermediate size polysaccharides devoid of xylose and mannose. Recent experiments in which it was possible to recover labeled exudate from stigma and stylar canal revealed that the high molecular weight component appears to be secreted in the region of the stigma while the intermediate molecular weight material is produced in the canal cells. A comparison of the chromatographic properties of these two exudates is presented in Fig. 6. It would appear that the stigmatic exduate gathered for injection studies was a mixture of the two.

In pollinated pistils, pollen tubes from newly germinated grains must pass through stigma-produced secretion product before encountering canal-produced material. Our data indicates that once the exudate enters the pollen tube, the more acidic, high molecular weight portion is utilized for tube wall formation with little or no breakdown while the less acidic, intermediate molecular weight fraction, rich in arabinogalactans, undergoes extensive breakdown, releasing monosaccharide and oligosaccharide fragments for re-assembly into the tube wall. It is this latter fraction that supplies precursor for conversion to glucose residues.

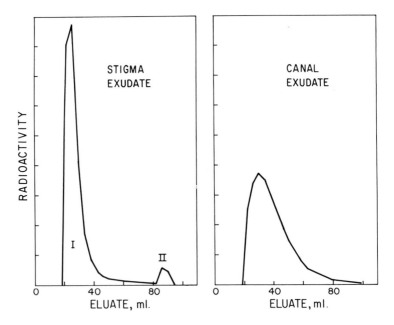

Fig. 6 Gel filtration profile on Sephadex
G-100 of stigmatic and canal exudates from a single
batch of detached *L. longiflorum*, cv. Ace pistils
which had been labeled 48 hr earlier with D-glucose-
1-^{14}C.

Exudate and Pollen Tube Formation in Pistils Receiving Labeled D-Glucose through the Vascular System

To examine such factors as rate of secretion
of stigmatic and canal exudate, relative contribution of each exudate to pollen tube wall formation,
and comparative composition of exudate and pollen
tubes at progressive stages of pollen tube growth
during cross- and self-pollination, an experiment
was devised to trace appearance of label in exudate
and pollen tube after labeling detached pistils
through the vascular system with D-glucose-U-^{14}C.
To minimize variations arising from sampling or

environmental factors, a sufficiently large number
of pistils were labeled at one time to provide complete
data from a single run (C. Labarca and F. Loewus,
submitted for publication).

At anthesis, 135 flowers were detached from
plants in a uniform stage of growth. Each was
stripped of tepals and stamens and placed in a vial.
The next day, each was given 0.1 ml of D-glucose-U-
^{14}C, 3 Ci/M. Two days after anthesis, groups of
45 pistils were cross-pollinated (Ace ♀ x Croft ♂)
or self-pollinated (Ace ♀ x Ace ♂) while the
remaining group of 45 pistils was left unpollinated.
Each day thereafter a group of 5 or 10 pistils was
removed for analysis of stigmatic and canal exduate,
and recovery of pollen tubes. To recover canal
exudate, stigma-style portions of pistils were
detached from ovaries and their canals flushed with
0.3 M pentaerythritol by means of a syringe.

Exudate production from stigma and canal,
measured as total carbohydrate per pistil, is
plotted in Fig. 7. At anthesis, only 0.1 mg of
carbohydrate per pistil was present as stigmatic
exudate. Within 24 hr, this had risen 10-fold.
In unpollinated pistils it continued to rise until
it reached 4 mg per pistil on the fifth day. Canal
exudate was produced at a lower rate and reached 2 mg
per pistil in the fifth day. Production of both
stigmatic and canal exudates was interrupted by
pollination.

Accompanying this increase in carbohydrate
secretion in unpollinated pistils was a rapid incorporation into exudate of ^{14}C-labeled products
derived from D-glucose-U-^{14}C as seen in Fig. 8. Three
days after label was given to the pistils, the
specific radioactivity of stigmatic exudate had risen
to 70,000 cpm per mg of carbohydrate. Again pollination interrupted the process and resulted in less
incorporation.

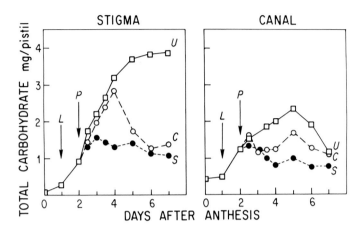

Fig. 7 Exudate production as measured by carbohydrate content in stigmatic and canal exduate of detached L. *longiflorum*, cv. Ace pistils. Plots are identified as U, unpollinated; C, cross-pollinated; and S, self-pollinated. Labeling, L, was initiated 24 hr after anthesis and pollination, P, 48 hr after anthesis.

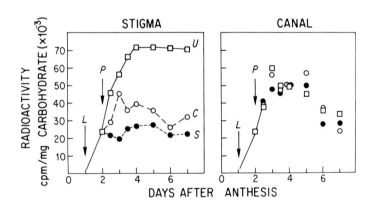

Fig. 8 Appearance of ^{14}C in stigmatic and canal exudates after labeling detached L. *longiflorum*, cv. Ace pistils. Same experiment as described in Fig. 7. See legend of Fig. 7 for abbreviations.

In contrast to the effect of pollination noted in stigmatic exudate, pollination did not influence the appearance of ^{14}C in canal exudate. All 3 groups of pistils, unpollinated, cross-pollinated and self-pollinated, reached a plateau of about 50,000 cpm per mg of carbohydrate in 3 days and remained at this level for 2 more days before declining as senescence altered the metabolic pattern.

As seen in Fig. 9, pollen tubes excised from

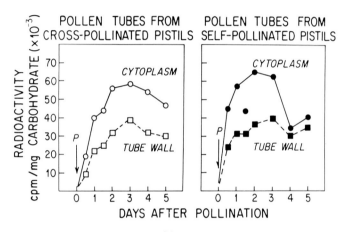

Fig. 9 Appearance of ^{14}C in pollen tube cytoplasm and pollen tube walls of detached cross- and self-pollinated L. *longiflorum*, cv. Ace pistils. Same experiment as described in Fig. 7 and 8. See legend of Fig. 7 for abbreviations.

these pistils exhibited a pattern of incorporation of label similar to that recorded for canal exudate. Tubes from cross-pollinated pistils reached the same level of incorporation found in canal exudate about 3 days after pollination, one day later than the exudate itself. Pollen tube cytoplasm exceeded canal exudate in specific radioactivity, probably reflecting the fact that pollen tubes penetrate highly labeled stigmatic exudate first, then become embedded in canal exudate of somewhat lower specific radioactivity.

Comparison of data in Fig. 9 with that in Fig. 8 provides a means of estimating exudate utilization as a source of carbohydrate for pollen tube wall biosynthesis. Two days after pollination, canal exudate contained 50,000 cpm per mg of carbohydrate, a minimal value since the method used to flush this secretion product from the canal resulted in low recovery when pollen tubes occluded the canal. Tubes recovered from these pistils contained cytoplasm with 60,000 cpm per mg of carbohydrate. Incorporation of label into wall polysaccharides lagged behind as one might expect of a wall assembly process dependent on delivery of carbohydrate precursor from the cytoplasm. Nevertheless, 3 days after pollination, wall polysaccharide contained 40,000 cpm per mg of carbohydrate. Clearly, most of the carbohydrate needed for tube wall formation was derived from pistil secretion product.

Conclusions

A mechansim does exist for utilization of pistil exudate by growing pollen tubes for tube wall formation. How exudate enters pollen tube cytoplasm is not clearly understood although ultrastructural studies offer possible clues (8). Nor is there as yet detailed information on the extent of polysaccharide fragmentation preceeding reutilization of secretion product for tube wall synthesis. But it can be said of pollen tube growth in L. Longiflorum that carbohydrate secretion by the pistil does have a major role in the fertilization process, supplying sufficient carbohydrate to the pollen tube to insure rapid biosynthesis of tube wall from grain to ovule and giving safe conduct to the generative nucleus in its journey down the pistil.

One cannot ignore the interesting possibility that for new cell wall formation, other processes such as the wound response, symbiosis, fruit development,

192

gall formation and secondary wall formation within
vascular tissues may depend on sources of secreted
carbohydrate which resembles the experimental system
described in this report.

References

1. W. G. Rosen in *Pollen: Development and Physiology* (J. Heslop-Harrison ed.) Butterworths, London (1971) p. 239.
2. C. Labarca, M. Kroh and F. Loewus, *Plant Physiol.* 46, 150 (1970).
3. C. Labarca and F. Loewus, *Plant Physiol.* 50, 7 (1972).
4. M. Kroh, H. Miki-Hirosige, W. Rosen and F. Loewus, *Plant Physiol.* 45, 92, (1970).
5. W. G. Rosen and H. R. Thomas, *Amer. J. Bot.* 57, 1108 (1970).
6. J. S. Pate and B. E. S. Gunning, *Annu. Rev. Plant Physiol.* 23, 173 (1972).
7. M. Kroh, C. Labarca and F. Loewus in *Pollen: Development and Physiology* (J. Heslop-Harrison, ed.) Butterworths, London (1971) p. 273.
8. W. G. Rosen in *Pollen: Development and Physiology* (J. Heslop-Harrison, ed.) Butterworths, London (1971) p. 177.

NATURE OF THE INTERCELLULAR SUBSTANCE OF STYLAR TRANSMITTING TISSUE

M. Kroh

Department of Botany
University of Nijmegen
Nijmegen, Nederland

Abstract

The intercellular material of stylar transmitting tissue is readily labeled with tritiated *myo*-inositol as can be demonstrated by electronmicroscopic autoradiography. It is soluble in warm water. Fractionation on Sephadex G-15 of the water extract of transmitting tissue from styles labeled with tritiated *myo*-inositol or tritiated glucose shows the bulk of label and carbohydrate to be present in one peak of low molecular weight material. This material appears to be a mixture of acidic carbohydrates. Acid hydrolysis releases, in addition to unidentified acidic oligosaccharides, free glucuronic acid and free galacturonic acid in a ratio of approximately 3:2. Whether the low molecular weight nature of the intercellular substance results from the extraction condition, or whether it occurs naturally remains to be determined.

Symbols

MI-2-^{3}H	=	*myo*-inositol-2-^{3}H
G-1-^{3}H	=	D-glucose-1-^{3}H
t.l.c.	=	thin layer chromatography
TFA	=	trifluoroacetic acid
EM	=	electronmicroscope

Introduction

Pollen tubes of most dicotyleous plants grow through the style in the intercellular substance of the stylar transmitting tissue. EM studies on pollinated styles suggest that pollen tubes utilize this intercellular material for elongation, possibly for cell wall synthesis (1). The intercellular material is assumed to be of pectic nature (1).

The present paper concerns preliminary studies on the nature of the intercellular substance. These studies are prerequisite for a later examination of the role of intercellular material for synthesis of pollen tube wall.

Materials and Methods

Flowers from *Petunia hybrida* from plants grown under greenhouse conditions were used. For each experiment, transmitting tissue was dissected from 200 styles of flowers one day before anthesis. The transmitting tissue was extracted in water for 1 hr at 39 C in presence of a trace of thymol, then removed by centrifugation, washed 3X with 2 ml of water and the combined extracts centrifuged 2X at 50,000g for 20 min.

For labeling experiments, water extract of transmitting tissue of 200 styles was mixed with that from transmitting tissue of 2 styles which has been labeled for 96 hr with 25 μC/style of MI-2-^{3}H or G-1-^{3}H. *myo*-Inositol is readily converted to uronosyl- and pentosyl units of pectic substances (2). During the period of labeling the length of the styles increased with an average of 9 mm. It should be noted here that the two labeled styles were detached 4 days prior to those pistils used to obtain transmitting tissue extract for carrier purposes. After water extraction, labeled transmitting tissue was ground in 1 ml of 80% ethanol, centrifuged, and the residue

washed 5 times with 2 ml of water. The radioactivity
of each fraction was determined.

Water extracts were fractionated on a column of
Sephadex G-15 (100 X 1 cm) and the column eluted
with 0.1 M acetic acid. Fractions of 1.5 ml were
collected at a rate of 6 ml/hr and analyzed for
radioactivity and/or total carbohydrate [phenol-
sulfuric acid (3)]. Fractions within the peak with
the highest amount of radioactivity and/or carbohydrate were pooled (designated G-15-III). Pooled
fractions from G-15-III were further separated by
t.l.c. on silica gel (DC-Fertigplatten, Kieselgur
F254, Merck) with indicated solvent mixtures* before
and after hydrolysis with 2 M TFA (sealed tube, 1 hr
120 C).

In some cases TFA-hydrolyzates were passed
through a column of Dowex-1 exchange resin (formate
form, 6 X 1 cm) prior to t.l.c. The acidic material
was eluted by washing the column successively with
0.1, 1.0, and 3.0 M formic acid.

Chromatograms were divided into areas corresponding to location of known sugars. These areas as
well as the origin and the area between origin and
galactose were eluted 3X for 1 hr with 1 ml of water.
The eluted areas of unlabeled material were analyzed
for total carbohydrate. The origin and the pre-
galactose area were also analyzed for uronic acids
with carbazole reagent. Labeled material of eluted
areas was counted in a Packard liquid scintillation
spectrometer at an efficiency of approximately 20% for
^{3}H.

---

*ethyl acetate-pyridine-water, 8:2:1, v/v for
neutral sugars (4), acetone- n-butanol-phosphate buffer (pH=4.6), 4:2:3, v/v for sugar acids (5).

EM autoradiographs were made from cross sections
of styles labeled with MI-2-^{3}H. Unlabeled styles were
studied electronmicroscopically before and after ex-
cising and extracting transmitting tissue with water.
Methods of labeling as well as EM techniques will be
described elsewhere (6).

Results

EM autoradiographs of MI-2-^{3}-H labeled styles
indicate that silver grains have accumulated above
intercellular substance and adjacent cell walls
(fig 1). One may assume that free MI was removed from
the labeled tissue during preparation of EM-auto-
radiography and that only that portion which was converted to carbohydrate units remained in intercellular
substance and cell wall material.

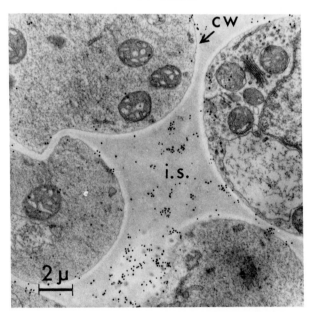

Fig. 1 EM-autoradiograph of a cross section
through the transmitting tissue of a style labeled with
MI-2-^{3}H. Note the accumulation of silver grains above
intercellular substance (i.s.) and adjacent cell walls
(cw).

A comparison of EM micrographs of unlabeled transmitting tissue before and after extraction with warm water shows that most of the intercellular substance of the transmitting tissue is removed (fig 2a, b). Corresponding to this, one finds 76–80% of the radioactivity from the labeled transmitting tissue in the water extract of this tissue before grinding (table 1).

TABLE 1

Distribution of Tritium among Fractions of Transmitting Tissue of MI-2-^{3}H and G-1-^{3}H Labeled Styles

Fraction	Source of label	
	myo-Inositol 2-^{3}H	D-Glucose- 1-^{3}H
H$_2$O-Extract before grinding	% 76	% 80
H$_2$O-Extract after grinding in 80 % Ethanol	14	10
Final Residue	10	10

199

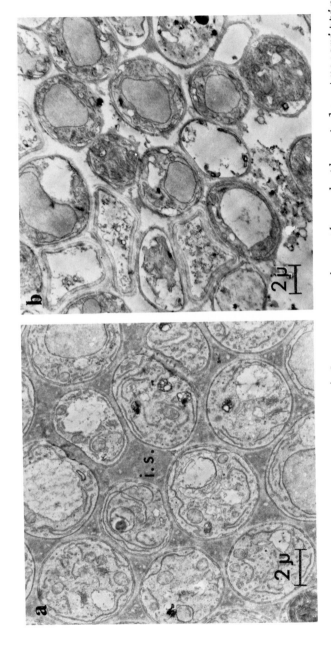

Fig. 2a, b. EM micrographs of cross sections through the stylar transmitting tissue before (a) and after (b) rinsing it with water.

Fractionation of the MI-2-^{3}H and G-1-^{3}H labeled extracts on Sephadex G-15 revealed that about 80% of the total radioactivity and total carbohydrate was present in a single peak (fig 3, G-15-III). In this peak 2.6 times more radioactivity was found in MI-2-^{3}H labeled material.

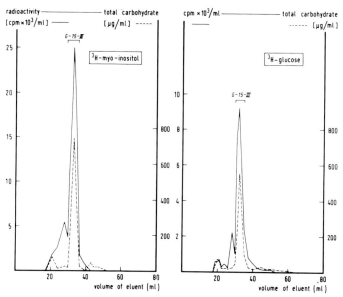

Fig. 3 Gel filtration of labeled water extract of transmitting tissue of Sepahdex G-15. Radioactivity (solid line), carbohydrate (dashed line).

Analysis of fraction III from Sephadex G-15 is summarized in table 2. The unlabeled material eluted from each area of the chromatogram was checked for carbohydrate. Results are given in the first column. About 24% of the total carbohydrate remained at the origin or between the origin and the region corresponding to galactose. The remaining carbohydrate had mobilities resembling hexoses. Of these, glucose was positively characterized with glucose oxidase. When unhydrolyzed G-15-III fraction from labeled tissue was separated by t.l.c. (column

2 and 4), very little radioactivity appeared in hexose areas. Most of it was present either in compounds of lower or higher mobilities. In the case of MI-2-^{3}H labeled G-15-III much of the pre-galactose/origin label was in acidic compounds. Hydrolysis with TFA followed by chromatography on Dowex-1-formate led removal of 31% of the label as acidic material and most of the remaining label still behaved on silica gel like the unhydrolyzed material (column 3). In the case of G-1-^{3}H labeled G-15-III a similar set of observations were obtained (column 5) but complicated by the fact that glucose also is a potential precursor of all hexoses and methylpentose residues encountered in G-15-III. The appearance of a considerable amount of unlabeled hexose in G-15-III may be a result of having used transmitting tissue that was excised and extracted directly after detachment of the pistils for carrier purposes. In contrast, labeled extract, containing only a small amount of label in hexose, was obtained from transmitting tissue excised 96 hr after detachment of the pistils. This matter is presently under study.

Efforts to characterize labeled components of G-15-III which remained on Dowex-1-formate or at the origin and in the pre-galactose area of the chromatogram after TFA hydrolysis (table 2) have not been completely successful. However, considerable information has been obtained that will assist in planning future experiments. The acidic material from Dowex-1-formate chromatography was a mixture of glucuronic acid, galacturonic acid and unidentified acidic oligosaccharides in which free uronic acid accounted for nearly one-half of the label. The ratio of free glucuronic acid to free galacturonic acid in these hydrolyzates was approximately 3:2. Possibly the ratio is even greater since an appreciable fraction of the glucuronic acid lactonized during acid hydrolysis and remained in the neutral fraction.

TABLE 2

Distribution of Tritium Among Eluted Areas of G–15–III (MI-2-^{3}H) and G–15–III (G-1-^{3}H) after t.l.c., before and after TFA Hydrolysis

Position on Silica Gel Thin Layer Chromatogram[+]	Unlabeled	myo – Inositol –2-^{3}H			Glucose –1-^{3}H	
	Non-Hydrolyzed	Non-Hydrolyzed	TFA[++] Hydrolyzed		Non-Hydrolyzed	TFA-Hydrolyzed
	Carbohydrate	radioactivity				
	%	%	%[+++]		%	%[+++]
Origin	24	20	36		25	16
Pre–Galactose		60	24		26	15
Galactose	21	4	2		5	8
Glucose	32	6	1		7	8
Mannose	21	1	0		7	3
Arabinose/Xylose/Fucose	2	6	4		18	13
Rhamnose	0	3	2		12	10
	100	100	69		100	73

+ Ethylacetate –Pyridine–H_2O, 8 : 2 : 1
+ + Trifluoroacetic Acid
+ + + Neutrals only

Further attempts to characterize labeled acidic G–15–III components of extract of transmitting tissue from MI-2-^{3}H labeled pistils are described in table 3. Here, material from the origin and pre-galactose region of chromatograms of unhydrolyzed G–15–III was recovered, hydrolyzed in 2M TFA and rechromatographed on silica gel plates in the same solvent system used to separate unhydrolyzed G–15–III. In the case of labeled material obtained from the origin of the previous chromatogram, about 60% behaved as acidic components and was removed by Dowex-1-formate treatment prior to t.l.c. Most of the remaining label was still present as slow moving oligosaccharide fragments. In the case of material from the pre--galactose region which was not run through Dowex-1-formate following TFA hydrolysis, most of the label remained near or on the origin when separated. These findings, although only preliminary, emphasize the

uronic acid and aldobiouronic acid nature of fragments released by TFA hydrolysis.

TABLE 3

Distribution of Tritium in Material Eluted from
the Origin and the Pre-galactose Region
of a Thin Layer Chromatogram after TFA–Hydrolysis

Position on Silica Gel Thin Layer Chromatogram+	G-15-Ⅲ myo-Inositol-2-^{3}H	
	Eluted Position after Thin Layer Chromatography+	
	Origin	Pre-Galactose
	TFA-++ Hydrolyzed	TFA- Hydrolyzed
	%+++	%
Origin	27	21
Pre-Galactose	4	60
Galactose	1	2
Glucose	0	0
Mannose	0	0
Arabinose / Xylose / Fucose	3	2
Rhamnose	5	15
	40	100

+ Ethylacetate-Pyridine-H_2O, 8:2:1
+ + Trifluoroacetic Acid
+ + + Neutrals only

Other experiments involving alkaline hydrylysis with warm 2% KOH or treatment with sodium borohydride followed by acid hydrolysis, support the conclusion that G-15-III contains a considerable amount of uronic acid of which the major portion is glucuronic acid.

In summary, intercellular material from transmitting tissue of *Petunia* styles does not consist of simple pectic substance. In appears to be a mixture of acidic, low molecular weight carbohydrates. Whether such a mixture occurs naturally in intercellular substance of transmitting tissue or whether it was produced by autohydrolysis during extraction with warm water is still undetermined and can only be clarified by future studies.

Acknowledgment

The suggestions and comments of Dr. F. Loewus and the skillful assistance of Mr. B. Kniuman are gratefully acknowledged.

References

1. J. van der Pluijm und H. G. Linskens, *Der Zuchter* 36, 220 (1966).
2. F. Loewus, *Ann. N.Y. Acad. Sci.* 165, 577 (1969).
3. M. Dubois, et al., *Anal. Chem.* 28, 350 (1956).
4. K. T. Williams and A. Bevenue, *J. Assoc. Offic. Agr. Chemists* 36, 969 (1953).
5. W. Ernst, *Anal. Chim. Acta* 40, 161 (1968).
6. M. Kroh and C. H. J. van Bakel, *Acta Bot. Neerl.* (in press).

THE ROLE OF THE GOLGI APPARATUS IN THE BIOSYNTHESIS
AND SECRETION OF A CELLULOSIC GLYCOPROTEIN IN
PLEUROCHRYSIS: A MODEL SYSTEM FOR THE
SYNTHESIS OF STRUCTURAL POLYSACCHARIDES

R. Malcolm Brown, Jr., [1] Werner Herth,[2]
W. Werner Franke,[3] and Dwight Romanovicz[4]

Abstract

The Golgi apparatus of *Pleurochrysis scherffelii*
functions in the biosynthesis and transportation of
the polysaccharide scales. The stepwise assembly
of the scale subcomponents has been followed with
the silver-methenamine-periodic acid technique as
well as normal poststaining procedures. The radial
microfibrils are first assembled in a bilayer of
parallel microfibrils, which later become unfolded.
Spiral bands of cellulosic microfibrils are then
deposited onto the distal surface. Finally, the
network of radial and "concentric" microfibrils
is covered with amorphous material, then the fully
assembled scale is secreted. The chemical composition of certain scale subcomponents is summarized in
this paper. The alkali-resistant "concentric" microfibrils consist of a cellulosic moiety intimately associated with peptides. Such an intimate association of structural polysaccharides with peptide
moieties also has been found for β-(1$\rightarrow$3)-xylan,

[1,4] Department of Botany, University of North
Carolina, Chapel Hill   27514
[2,3]Division of Cell Biology, University of
Freiburg, Freiburg i. Br. West Germany.

β-(1→4)-mannan, β-(1→3)-glucan, and even cotton cellulose. This evidence suggests a general similarity of the pathways of animal glycoprotein biosynthesis and plant cell wall polysaccharides including structural polysaccharides. This general pathway derived from the *Pleurochrysis* model has been compared with the more classical pathway postulated for cellulose biosynthesis in higher plants.

Symbols

PA	= periodic acid
ER	= endoplasmic reticulum
"Z"	= a stage of cisternal membrane unfolding
GLC	= gas-liquid chromatography
TMS derivatives	= trimethylsilyl derivatives
NMR	= nuclear magnetic resonance spectoscopy
DP	= degree of polymerization

Introduction

In recent years, while considerable attention has been devoted to cell wall and polysaccharide biochemistry, relatively little is known about these biosynthetic pathways in terms of cellular substructure and function. In general, the plant cell wall has eluded a precise descriptive study of cytological events leading to its formation, primarily because it is composed of a rather heterogeneous and amorphous assemblage of microfibrillar and matrix substances. While certain of these constituents have been shown to be produced by the Golgi apparatus (25) little is known about the biosynthetic pathway of structural polysaccharides*, especially cellulosic

---

*For definition of "structural polysaccharides" see Herth, et al., (13).

products. Recent biochemical evidence by Ray (30) has suggested that a subcellular fraction with marked β-(1→4)-glucan synthetase activity is confined largely to membranes of the Golgi fraction. While the role of the Golgi apparatus in cellulose biosynthesis also is suggestive for higher plants from indirect evidence, e.g., hypertrophy of Golgi vesicle cells with increased wall formation, little information concerning the secretory pathway is known with the exception of the recent work by Brown and co-workers (2, 3, 4). In this instance, a unicellular marine haptophycean alga has been used as a model system to investigate cellulose biosynthesis. On the other hand, more details are known about cellulose biosynthesis in certain prokaryotic systems, namely, *Acetobacter* (8).

It is well known that marine haptophycean algae are capable of producing an architectually defined scale (17, 18, 29). The discovery that Golgi--produced scales contain a cellulosic moiety came about in a somewhat strange manner. The early investigations of Green and Jennings (11) failed to reveal the presence of cellulose, but their findings did confirm other polysaccharides.

The investigations of Brown and co-workers (3, 4) showed that an alkali-stable moiety of purified scales could be hydrolyzed to glucose monomers and that by mild hydrolysis, cellobiose could be obtained (13). In addition, a spectrum of other biochemical, biophysical and cytological techniques have confirmed that a cellulosic moiety exists in the scales of *Pleurochrysis scherffelii*. These tests included analysis of linkages of the sugar monomers by means of nuclear magnetic resonance spectroscopy, x-ray diffraction, and various chemical tests (3, 4, 13).

Recently, it has been suggested that the harsh alkali-resistant cellulosic scale moiety is covalently linked with a peptide moiety (13). This evidence has led to the concept that *Pleurochrysis* cellulose exists in a form of glycoprotein. Adding to this growing list of analyses of scale products, we are accumulating cytological and cytochemical evidence which describes the specific assembly stages of various scale subcomponents by the Golgi apparatus.

The purpose of this presentation is to consider the cytological, cytochemical and biochemical evidence for the role of the Golgi apparatus in the production of structural polysaccharides. Because *Pleurochrysis* is such a distinctive system in which the architecturally-defined Golgi polysaccharide products can be isolated and analyzed, there is an opportunity to correlate this background of information with data from other polysaccharide synthesizing systems, namely the structural polysaccharides components of *Acetabularia*, *Caulerpa*, lily pollen tube wall, cotton, and onion root, as they relate to growth and differentiation of the plant cell in general.

Materials and Methods

Cultures. Axenic cultures of *Pleurochrysis scherffelii* were grown on agar plates in an enriched van Stosch sea water medium as previously described (4).

Isolation of scales. Scale fraction A was obtained according to the methods of earlier described (4, 13).

Extractions. For the details of organic, chlorite and alkali extractions see text.

Electron microscopy. We used the same fixation and embedding conditions as described in Brown, et al.

(4). Sectioned material to be stained with silver
hexamine (28) was mounted on formvar-coated stain-
ess steel grids which were then treated to block or
reveal certain staining moieties. (For details, see
text and Romanovicz, et al., in prep.) Negative
staining was performed by adding 2% PTA (adjusted
to pH 6.8 with 1N KOH) to specimens which had been
first dried on carbon-stabilized, formvar films,
then briefly rinsed in distilled water. Specimens
were examined with a Hitachi Hn-11E electron microscope.

Chemical characterization. For details see
Brown et al., (4) and Herth et al., (13).

Staining reactions. See Brown, et al., (4) for
details of the zinc—chloride-iodine staining.
Ruthenium red staining was performed with a 1%
aqueous solution. Alcian blue staining was used
according to the methods of Parker and Diboll (27).

Results

I. Cytological and Cytochemical Observations

A. Cell Wall Structure

Functionally, the cell wall of *Pleurochrysis*
probably differs little from higher plant cell walls,
in that it provides a structural framework which
encompasses the cell protoplasm. This, in turn,
probably determines ultimate cell shape and may regulate cell growth. The cell wall structure of
Pleurochrysis however, differs significantly from
higher plant cell wall systems in that it is composed
of a morphological unit known as the scale. The unit
scale is well known for other haptophycean cells (17)
but in most of these algae, the scales are either
released into the culture medium, or form a rather
loose layer over the cell surface. In most Haptophyceae the predominent phase consists of a motile

zoospore. In *Pleurochrysis*, however, the major phase
is the non-motile vegetative cell which produces a
very compact and definitive cell wall by aggregation
or stratification of the unit scales.

Light microscopic examination of *Pleurochrysis*
cell walls reveals a positive birefringence. The
walls also exhibit a positive iodide dichroism (2).
Furthermore, the walls of *Pleurochrysis* react with
the periodic acid Schiff's test, indicating that it
is carbohydrate. This is substantiated at the ultrastructural level by silver methenamine following
periodate oxidation (Fig. 8). *Pleurochrysis* cell
walls also react with Alcian stains. At pH 0.5,
positive Alcian staining results under conditions
where carboxyl groups are fully protonated. Following methylation, the cell walls fail to react with
Alcian stains at pH 0.5 or pH 2.5. Saponification
with KOH reestablishes carboxyl groups which subsequently react strongly with Alcian Blue at pH 2.5,
but negatively at pH 0.5. The protonation and methylation reactions suggest that at least one constituent of the scale cell wall is a sulfated polysaccharide (15). These light microscopic reactions are
confirmed at the ultrastructural level with the
silver methenamine reaction using specific blocking
reagents such as iodoacetate for sulfate groups or
bisulfite for aldehyde groups (Table I). These
blocking reagents have been used in combination
with periodate oxidation in order to distinguish the
free aldehydes present from those that were produced
by the specific oxidation of 1-2 glycol groups in
the polysaccharides. Treatment of vegetative cells
with ruthenium red reveals a positive staining of
the cell wall layers suggesting acidic constituents
of the cell wall (5).

B. Scale Structure

When vegetative cells of *Pleurochrysis* undergo
zoospore formation, the cell wall becomes dissociat-

PLANT CELL WALL POLYSACCHARIDES

Table 1

Staining of _Pleurochrysis_ vegetative cells with silver methenamine

Treatment	Chemical reaction	Silver deposition			Preliminary interpretations
		Wall Scales	Intra-cisternal Scales	Membranes	
(a) Ag	reduction of Ag^+ at cellular sites	+	+	+	general cytoplasmic staining but more intense with wall and intra-cisternal scales.
(b) Iodoacetate-Ag	blockage of reduced sulfur by acetylation	-	-	(+)	Scale silver deposition from untreated controls due to sulfhydryl groups.
(c) Iodoacetate-PA-Ag	reduced sulfur blocked by acetylation and periodic acid oxidation of 1-2 glycols to aldehydes	+	+	(+)	restoration of silver deposition because of PA oxidation of 1-2 glycols to free aldehydes. Sulfhydryl blockage remains.
(d) Iodoacetate-PA-HSO$_3$ Ag	blockage of reduced sulfur by acetylation, production of free aldehydes by PA, followed by blockage of these aldehydes by HSO$_3$.	-	-	(+)	loss of silver deposition due to blockage of PA-oxidized free aldehydes.

ed into the unit scale structures. They can also be removed from whole vegetative cells by sonication (Herth, unpublished) and by high pressure decompression (Brown, unpublished data). Differential contrifugation methods can be used to purify and enrich scale fractions from protoplasmic and whole cell contaminants. The unit scale consists of radial and "concentric" elements of microfibrillar nature. Upon this microfibrillar base can be deposited other organic (e.g., pectic) or inorganic constituents, for example, the calcium carbonate rims of the coccoliths (10). The radial and "concentric" microfibrils are readily observed in negatively stained isolated scale fractions (Fig. 4). The radial patterns exhibit a very precise symmetry in two planes. Amorphous, acidic sulfated polysaccharide components can be deposited in between the radial microfibrils (Fig. 3).

The radial microfibrils have about the same dimensions as the "concentric" microfibrils; but they differ in their cytochemical reaction to the silver methenamine reaction and their stability in alkali washes. Furthermore, the radial microfibrils are removed by pectinase whereas the "concentric" microfibrils are only degraded by cellulase (Herth and Brown, unpublished data). Evidence to suggest degradation of radial microfibrils rather than separation from the "concentric" microfibrils is based on two points: (a) Following 5% KOH treatment which removes the radial microfibrils (Fig. 5), the supernatants of these preparations do not contain fibrillar components that could be detected (by negative staining) in the pellet after high speed centrifugation (48,000 X G for 45 min.). Instead, the dissolved polysaccharides can be precipitated with ethanol and CaCl$_2$ ions (Herth and Ramonovicz, unpublished data); (b) Different stages of radial microfibrillar dissolution could be detected with negative staining after treatment with pectinase or dilute alkali.

The so-called "concentric" microfibrils originally described to be concentrically organized (4)
are now known to be arranged in several continuous
spiral bands of approximately 8 to 10 microfibrils
each. Because these microfibrils are so closely
packed, it was not possible to deduce the spiral
configuration until partially alkali-degraded scales
were observed (Fig. 6). The individual spiraling
microfibril consists of a ribbon which is approximately 40 Å wide and approximately 20 Å in the
narrow axis which lies adjacent to the radial microfibrils and parallel to the flat plane of the scale.
The microfibril unit has subcomponent structures as
shown in Fig. 7 in which a single sprial microfibril
has split into at least three linear units each
which may actually be associations of several glucan
polymers. In this state, the spiraling microfibrils
are structurally more flexible and the scale tends
to fold, or the "concentric" bands tend to separate
from one another. The resulting physical damage
produces consistent breakage loci in the spiral
microfibrils, similar to those that occur in the
central region or "eye" of the unaltered scale.
These latter natural breakage loci are thought to
occur during the final stages of spiral microfibril
deposition when the limited secretion space causes
the bending angle to increase beyond the maximum
flexibility of 2% for the crystalline product.

C. Structure of Protoplasmic Organelles
and Constituents with Particular Reference to
the Golgi Apparatus

Figures 1 and 8 depict a typical view of the
protoplasmic contents of a vegetative cell of
Pleurochrysis. The uninucleate cell contains two
parietal chloroplasts. The outer nuclear membrane
is continuous with and surrounds both chloroplasts.
An extension of the outer nuclear membrane forms an
amplexus which becomes the site of vesicle delimitation for the forming face of the Golgi apparatus

(Fig. 1).

The single Golgi apparatus of *Pleurochrysis* is a
most complicated organelle and is composed of a
stack of differentiating cisternal membranes. The
Golgi apparatus is polarized, with the secreting
face of inflated cisternae always oriented toward
the cell surface (Fig. 2). The forming face of the
Golgi apparatus apparently becomes organized with
the fusion or coalescence of vesicles which seem to
originate from the budding face of the amplexus.
This particular region of the amplexus is free of
ribosomes. The forming face of the Golgi apparatus
consists of thin, very closely appressed cisternae
of rather uniform dimensions. Midway through the
differentiating stack of cisternae "central dilations"
appear (9) (Fig. 2). These dilations are thought to
contain the precursors and enzymes for the synthesis
of polysaccharides because all cisternae on the forming side of these dilations have no visible product
within them, while all cisternae on the secretion
side of these dilations contain a visible secretory
product. The cisternae of the secreting face of the
Golgi apparatus are considerably larger than those
of the forming face (Fig. 2). This may be a result
of a secondary fusion of vesicles migrating from the
ER to the periphery of the Golgi apparatus. These
vesicles appear to carry additional amorphous
polysaccharide products to the surface of the microfibrillar network which has been produced in earlier
stages.

Transport of the mature scales to the cell surface occurs when the cisternal membranes fuse with
the plasma membrane, releasing the scale products.
Two bundles of microtubules flank the secretory
face of the Golgi apparatus (6). Apparently a function of these microtubules may be to continually
orient the secreting face of the Golgi apparatus so
that the products can be transported to the cell
surface. Other functions have been postulated for

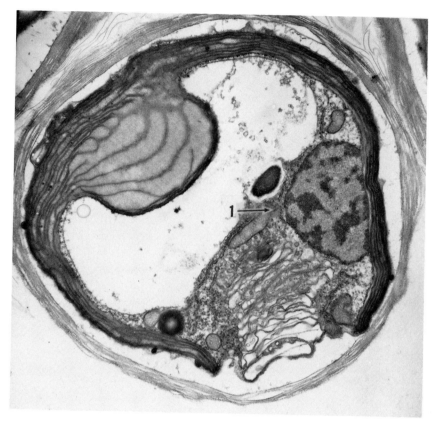

Fig. 1 Survey electron micrograph of a median
optical section through a mature vegetative cell
(normal post staining with lead citrate and uranyl
acetate). Note the prominent Golgi apparatus in
the lower right hand corner surrounded by endoplasmic
reticulum and many polysomes. Note the amplexus or
extension of endoplasmic reticulum from the outer
nuclear membrane (arrow 1), note also, prominent
mitochondria associated with the Golgi region. The
Golgi apparatus is polarized and the secreting face
always is directed toward the cell surface. Note
the parietal chloroplast with protruding pyrenoid
and the wall of stratified scales. X 10,000.

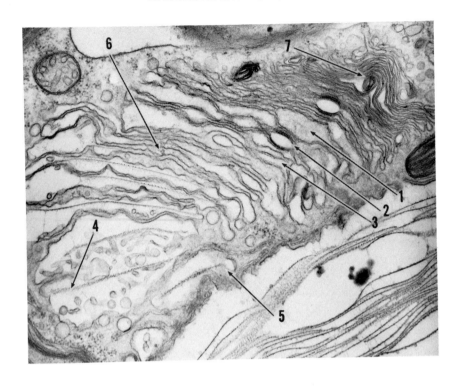

Fig. 2   The Golgi apparatus of *Pleurochrysis*
shown in median section (normal post staining with
lead citrate and uranyl acetate). Note the differentiating stack of cisternae beginning with compact
cisternal development adjacent to the amplexus
upper right hand corner) and progressing toward the
cell surface with inflation. Two types of central
dilations are seen in the proximal region of the
Golgi apparatus: arrow 1 depicts the large dilation
with visible product and arrow 2 shows the smaller
electron transparent dilation that is surrounded
by heavy electron-dense staining of material next to
the membrane. The first visible scale elements
(radial microfibrils) appear as a double layer in
the cisterna at arrow 3. Mature scales are shown
at arrow 4 where amorphous material coats the microfibrillar scale network. Scale secretion occurs at
arrow 5 where the distal-most cisterna fuses with
the plasma membrane and releases the scale to the
cell surface where it stratifies to form the cell
wall. Arrow 6 depicts a cluster of polysomes in
between the cisternal stacks probably involved in a
specific stage of scale assembly. Whirls of membrane
(arrow 7) appear in the forming region of the Golgi
apparatus. Polysomes and vesicles appear along the
cisternae of varying stages of differentiation along
the cisternal stack. X 24,000.

these structures (6).   When cells of *Pleurochrysis*
are treated with colchicine, the microtubular
bundles disappear and the secretory face becomes disorganized (Brown & Mclean, unpublished).   The continuing intracellular production of scales results in
autophagic comsumption of scales since the secretion
mechanism appears to be inhibited (Fig. 18).

Adjacent and interior to the plasma membrane
is a large peripheral sac which is continuous over
the cell surface except for the region immediately
opposite the secretion face of the Golgi apparatus.
Through this "hole", scales are secreted.   This
parietal cisterna is thought to represent the "ball-
bearings" or flexible surface upon which the proto-
plast could rotate to ensure a uniform deposition of
scales on the cell surface (4, 5).

D. Developmental Pathway of Scale Biogenesis (see Fig. 15)

Fig. 15 depicts our model of the secretory
pathway of scale production by the Golgi apparatus
in *Pleurochrysis*.   This model has been developed
from the information of several thousand electron
micrographs and dozens of fixations during the past
several years.   Other methods used to arrive at this
model include serial sections, freeze etch replicas,
negative staining, and the electron microscope tilt-
ing stage for stereoscopic analyses.   Because a
scale is produced about once every two minutes, the
differentiation between one cisterna and the next
may be quite transient (2, 5).   Thus, all structural
variations that occur within this time usually are
not present in a single electron micrograph.   With
these data at hand, we have been able to suggest a
more precise developmental pathway than heretofore
possible (for comparison with the original pathway
proposal for its modification, see Brown, et. al., 4).

Cisternal Membrane Delimitation. The immature forming face of the Golgi apparatus consists of a coalescence of vesicle membranes into a flat sac (Fig. 13). As this sac grows, presumably by continued coalescence of these budding vesicles from the ER, certain regions differentiate within the cisternal membrane. These areas include incipient organizing centers for the polymerization and/or crystallization of scale polysaccharide products. Staining with silver methenamine reveals two types of dilated organizing centers; one which is filled with product, (i.e., reactive with silver), and the other which remains non-reactive with silver under the conditions employed (Fig. 8, 9). The non-staining organizing center is believed to contain the precursor monomers, enzymes, and/or noncrystalline polymers which will become organized into the "concentric" microfibrils of cellulosic glycoprotein. The silver-reacting dilated organizing center contains the precursors and enzyme machinery for the synthesis of radial microfibrils (Fig. 11).

Formation of Radial Microfibrils. The **radial** microfibrils of the scale exhibit a bilateral symmetry in two planes (Fig. 3). The radial microfibrils are formed adjacent to the inner cisternal membranes in a folded, mirror-image pattern, i.e., the long axis of symmetry is folded when the radials are deposited.

The radial microfibril precursors of the silver reacting dilation seem to be fed mostly into a central compressed region of the cisternal membranes (Fig. 11), although the precursors can occasionally be seen feeding toward the periphery of the Golgi cisterna (Fig. 14). The folded scale axis has been observed quite frequently (see Figs. 10, 11, 12, 13). In addition, this stage has been proven with stereoscopic observation of tilted specimens. The individual radial microfibrils in the folded configuration are arranged predominantly parallel to

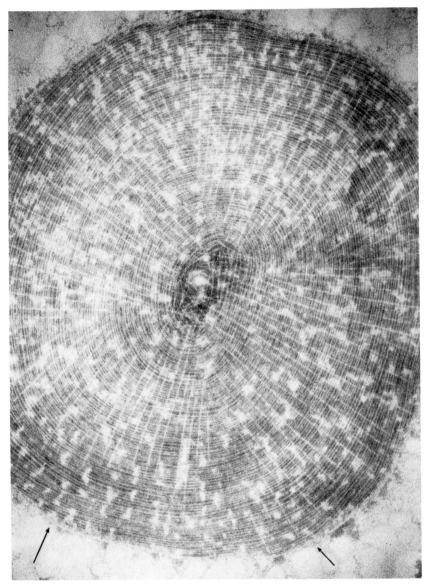

Fig. 3 Scale from vegetative cell with amorphous material in between the radial and "concentric"
microfibrils. Note the two axes of symmetry for the
radial network. The "concentric" microfibrils appear
to spiral and some of them end on the periphery at
arrows. X 84,000.

the long axis of symmetry. Frequent cross sections perpendicular to the long axis reveal many parallel and evenly spaced microfibrils (Figs. 10, 11). To better understand the parallel compression stage of the radials along the long axis, a negatively stained, mature circular scale was projected onto an enlarging easel and the radial microfibrils traced (Fig. 16A). In this instance, there are approximately 25 microfibrils per quadrant. If the individual microfibrils of one bilateral axis are compressed into an elliptical space without altering their length or order of arrangement, the earlier stages of development, especially the parallel compression stage can be visualized (Fig. 16B). Similar patterns have been seen in grazing sections of early scale development in the Golgi apparatus and confirm not only the folded configuration of radials, but also the compression of folded radials along the long bilateral axis of symmetry.

One of the most remarkable features in scale development is the unfolding process that must take place in order to produce the ellipsoidal or circular scale. Fig. 7 depicts this unfolding process, in which a central compact region of the Golgi cisterna containing the folded radial microfibrils goes into a "Z" formation*. The forces for this particular membrane transformation and movement are unknown. Nevertheless, such movement could account for a logical sequence for the actual unfolding of a scale shown in Fig. 8. Once the scale unfolds, the radial microfibrils at first remain in the compressed phase, the two scale halves still separated and at a slight angle to one another (Fig. 13, arrow 1). Presumably, shortly after this stage the radial

---

*Most of the evidence now speaks for a single unfolding step, but we cannot exclude the possibility of a sequential double unfolding to produce the four quadrants of radial symmetry.

microfibrils separate from their attached sites on
the membrane and spread into a planar bi-radial configuration which produces the circular or slightly
elliptical base onto which cellulosic microfibrils
subsequently will be deposited.

Development of "Concentric" Microfibrils. Once
the radial network is synthesized, unfolded and
spread, the empty dilation that earlier contained
the precursor of the radials, migrates to the periphery of the cisterna where it later undergoes ab-
scission from the parent cisterna. The empty radial
precursor sac is always situated on the proximal
side of the cisternal membrane (Fig. 10). Corres-
pondingly, on the distal side, a very narrow constriction appears attached adjacent to the exact geo-
metrical center of the radial network. Other tubules
seem to associate with the periphery of the distal
side of the cisterna. In this stage, the radial
microfibrils are now enclosed by a very compact cisternal membrane (Fig. 10, 11, 13, 14). The distal
associated tubules are involved in the synthesis
and/or transport of the "concentric" microfibrils to
the distal side of the radial network. Since the
so-called ""concentric" microfibrils consist of a
spiral arrangement of four bands* of normally eight
to ten continuous, mutually attached, parallel microfibrils per band, these must be polymerized, crystallized, and cross-linked by a very well organized,

---

*This is the normal case for mature vegetative
scales. Smaller scales produced by zoospores or
those produced after treatment with various inhibitors have fewer bands (generally 1-2), each with
fewer turns of the spiral. The number of microfibrils per band, however, remains relatively constant.

rather complex process resulting in this special arrangment. We cannot yet decide between two alternative possible processes of spiral band formation and arrangement:

a)   Perhaps we have a Spinnerette-type functioning of a compound polymerization center in the central constriction (see Figs. 14, 15). According to this hypothesis, the four bands of eight to ten individual microfibrils are *either* polymerized, cross--linked by peptides and crystallized in one continuous process, *or* cross-linked and crystallized from previously polymerized glucan chains. Then they are fed through the central tubule into the very thin cisternal lumen onto the distal side of the radial network where the growing bands are pushed to the periphery. Final attachment of the spiral bands to the radial network occurs at the periphery of the radials where the four bands are compressed together adjacent to the peripherial cisternal membranes.

b)   The polymerization, crystallization, and cross--linking of the four bands could also start simultaneously from four starting points at the rim of the scale. The growing tips of the parallel bands would then move inwardly. In this case the polymerization center must rotate on an axis, possibly the axis of the scale.

In both cases, the cisternal membrane limitation restricts the movement of the microfibrils so that they can occur only toward the sac periphery according to possibility (a) and toward the center in case (b). In both alternatives, the bands of parallel microfibrils are synthesized continuously, laid down first at the periphery of the cisternae, and spiral inwardly until finally the space restriction causes microfibril fracture, or disordered synthesis in the central region. The movements leading to the 4-5 turns of the spiraling bands seem to be more easily explained with rotating movements of the feeding tubules from the cisternal periphery towards the center of the scale while fibrils are formed and laid down according to hypothesis (b), thus making this possibility a little more attractive. But perhaps the tendency of the crystalline, parallel microfibrils not to be bent more than necessary in a restricted space provides enough force to ensure the spiraling, similar to that of a clockspring. That some force is used to lay down the spirals is clearly visible from the bending of all radials in the direction opposite the spiraling* due to the inherent unwinding tendency of the crystalline cellulosic microfibrils (Fig. 4).

With the addition of the spiraling microfibrils, the basic microfibrillar skeleton of the scale is laid down and cemented together.

Addition of Other Polysaccharides to the Microfibrillar Skeleton. As the fully-formed scale microfibrillar skeleton moves in the differentiating cisternal membranes, other vesicles, or perhaps remains of either radial or cellulosic precursor vesicle membranes participate in the addition of other polysaccharides to the surface of the microfibrillar skeleton. This begins with addition on the scale periphery, and is followed by a centripetal deposition of product, until finally the entire scale is coated (Fig. 10). Once these polysaccharides are added, the scale no longer is held in restricted space by cisternal membranes. Still, the scales cannot turn around within the membranes, and are thus secreted to the periphery in the same orientation that they were formed. In the case

---

*Inward-to-outward spiraling in a counter--clockwise direction seen from the distal surface of the radials.

with *Pleurochrysis*, the addition of amorphous polysaccharides to the scale microfibrillar skeleton
results in the mutual adhesion of scales to form
the compact vegetative cell wall. In zoospores,
scales are synthesized, but without the amorphous
polysaccharides, and thus they tend to separate from
the cell surface soon after they are secreted.

  Scale Secretion Mechanisms. Scales are
brought to the cell surface by a fusion of the cisternal membranes with the plasma membrane. Presumably, there is some membrane re-cycling, because
the total cell surface membrane would greatly ex-
pand if total cisternal membranes were incorporated
into the plasma membrane. Evidence seems to indicate that many small vesicular membrane fragments
which occur adjacent to the plasma membrane may be
part of the re-cycling of cisternal membranes (2).
Electron-dense autophagic vacuolar structures may
be involved in this process, but at present this
stage has not been thoroughly investigated. It is
important to note that on either side of the secreting face of the Golgi apparatus there exists
a single bundle of approximately 200 microtubules.
These microtubules exist in a hexagonal array and
have been shown in detail with freeze-etching
(4) and other methods. Recent experimental data
from colchicine treatments (Brown and McLean, in
preparation) have shown that this drug depolymerizes
the microtubular system, as expected. However, this
seems to alter the direction of secretion so that
scales are unable to be secreted to the exterior.
Instead, they accumulate inside the cell, and the
autophagic vacuolar system becomes activated and
engulfs the scales as rapidly as they are produced.
Sometimes, entire cisternal stacks are digested
(Fig. 18). Thus, it appears that at least one para-
meter of the secretory regulatory mechanism resides
in the microtubular bundles, and apparently the
autophagic system is a regulatory mechanism for
stopping scale secretion and/or re-cycling some of

the components. Preliminary results of Cytochalasin B treatment also suggest the involvement of microfilaments in some of the processes leading to scale formation and secretion (Herth & Brown, unpublished data, see also Herth, et al., 12).

Protoplasmic Movements and Scale Secretion. Because *Pleurochrysis* cells have only a single Golgi apparatus, and because the scales are evenly distributed over the cell surface, a mechanism must exist to provide a random directional movement of the secreting phase relative to the surface to insure a uniform deposition of scales. This was first hypothesized by Brown (2) from electron micrographs and later reconfirmed by time-lapse cinematography (5). The remarkable cells of *Pleurochrysis* undergo a continuous protoplasmic rotation, which apparently insures uniform distribution of the pre-formed scale cell wall products. Not only is this peculiar movement found in *Pleurochrysis* but in all members examined so far which produce a preformed scale product. This includes not only the Haptophyceae, but also the Prasinophyceae.

At first it was thought that the microtubular bundle might be involved in this protoplasmic movement, but it is now believed that another force is responsible for this movement, one which may reside in finger-like projections of the cisternal sac which surrounds the Golgi apparatus and which encompasses the cell (Fig. 17). With freeze-etching, the surface of the finger is reminiscent of contractile protein configurations found in bacteriophages (Brown, unpublished data). Freeze-etching also shows that the fingers can exist in two states, a spherical configuration without much degree of order and an elongated cylinder which has a high degree of order. Presumably, a transformation between spherical and cylindrical states of many of these structures may account for the gradual and very slow protoplasmic rotation observed. It should be noted that protoplasmic rotation is much slower than cyclosis and cannot be adequately observed without time-lapse techniques.

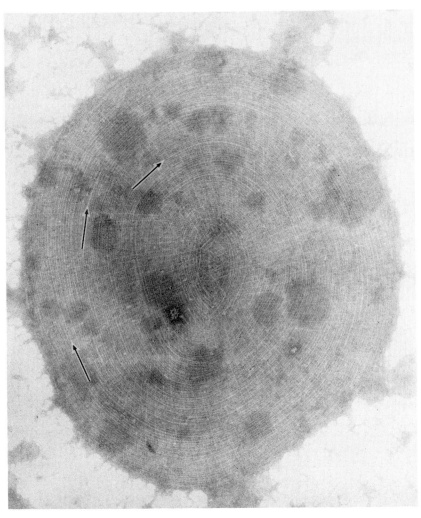

Fig. 4   Negatively stained scale from vegetative cell of *Pleurochrysis* with amorphous material removed. Note the ordered radial and "concentric" network of the microfibrils. Spiraling of the "concentric" network can be detected by occasional discontinuity of the spiraling bands (arrows). Note the slight curvature of radials in the counterclockwise direction of the spiral which is suggestive of the force of direction of the assembly of the spiral network. X 59,000.

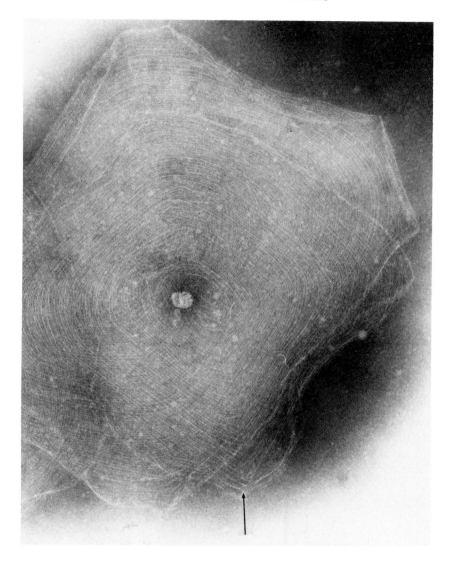

Fig. 5 Scales from vegetative cells of
Pleurochrysis after a short treatment with 5% KOH.
Note the removal of the amorphous substances and
radial microfibrils. The remaining "concentric"
microfibrils have less structural stability and
therefore the rims tend to fold back on one another.
In addition, numerous breakage loci throughout the
spiral network can be seen (arrow). X 68,000.

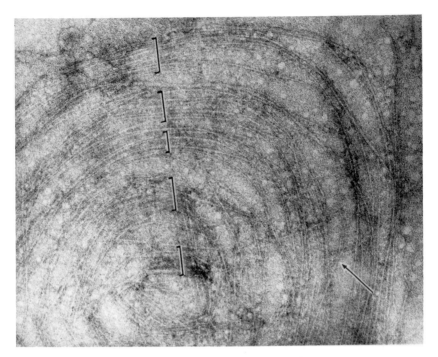

Fig. 6 A more advanced stage of degradation
of scales of vegetative cells of *Pleurochrysis* treated
with 5% KOH. Note again the complete absence of
radial microfibrils. Here, the spiraling network
of microfibrils is beginning to unravel. Note the
distinct bands of spiral microfibrils shown in between the brackets. Between 5 and 9 microfibrils
comprise the band. X 92,000.

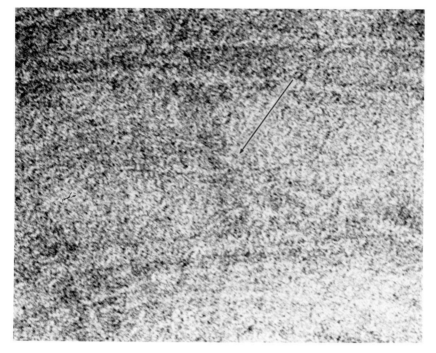

Fig. 7   Details of region in Figure 6 indicated
by arrow showing the sub-unit structure of the
spiral microfibrils. A single microfibril, arrow,
breaks down into two or three subcomponents.
X 440,000.

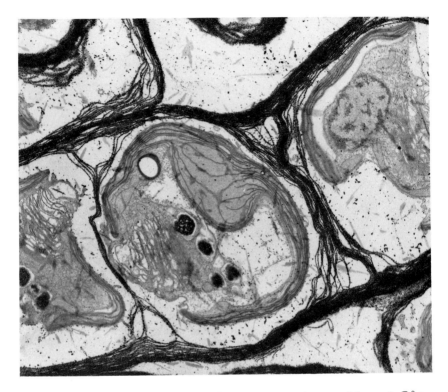

Fig. 8 Actively growing vegetative cells of *Pleurochrysis* in the filamentous phase. Note the electron dense wall layers resulting from silver deposition (silver methenamine method with periodate oxidation). Each vegetative cell contains a Golgi apparatus in the section plane as well as a single parietal chloroplast with bulging **pyrenoid**. In the right cell the nucleus is prominent and lies adjacent to the Golgi apparatus. Note the newly formed cross wall between the two vegetative cells on the left. The central cell exhibits a large conspicuous vacuole with several electron-dense autophagic vacuoles lying between the Golgi apparatus and the large vacuole. X 8,000.

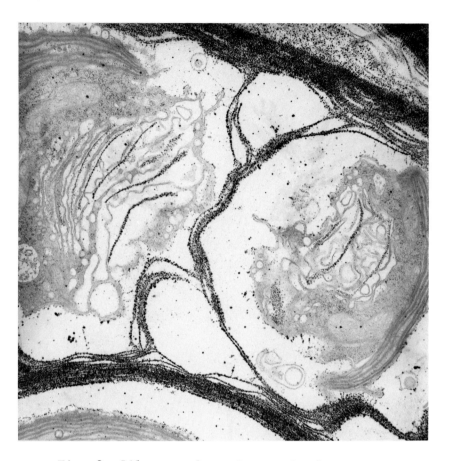

Fig. 9 Silver methenamine-stained preparation of two actively growing vegetative cells. Note the prominent Golgi apparatus in each cell and the identity of the staining properties of the Golgi scales with that of the wall material. X 13,000.

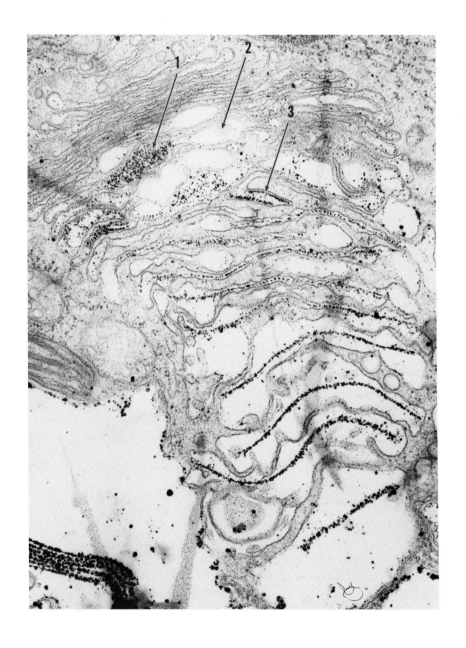

Fig. 10 Over-all view of a Golgi apparatus stained with PA-silver, showing the dilations of the radial precursor pools which react heavily with silver (arrow 1), and the cellulosic precursor pools which are electron transparent (arrow 2). The amplexus and budding ER is shown in the upper left hand corner. A stage of folded radial microfibrils is shown at arrow 3. The next distal-most cisterna contains a folded scale with radials but without added concentrics. The distal scales react more heavily with silver, indicating the addition of amorphous polysaccharide. Note cell wall in the lower left hand corner. X 41,000.

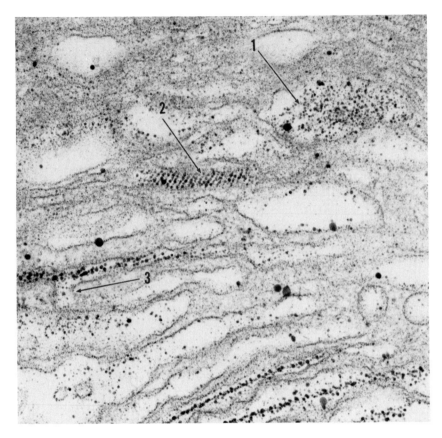

Fig. 11 Details of the radial precursor pool
(arrow 1) and the folded radial microfibrillar network (arrow 2) as shown by the silver technique.
Note that the folded radial microfibrils are compressed parallel to one another. The cisterna at arrow
3 contains an unfolded radial network with a central
feeding tubule below it. The cisternae in the lower
right hand corner contain fully assembled scales
with amorphous product. X 71,000.

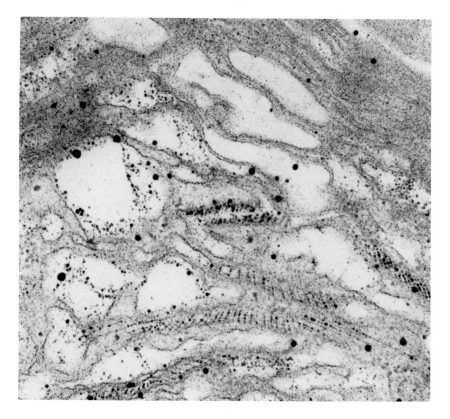

Fig. 12   Detail of the "Z-stage" radial unfolding in the Golgi apparatus as shown by the silver technique.   Compare with stage in schematic diagram. Also compare with Figure 13 which shows a later stage of unfolding. X 58,000.

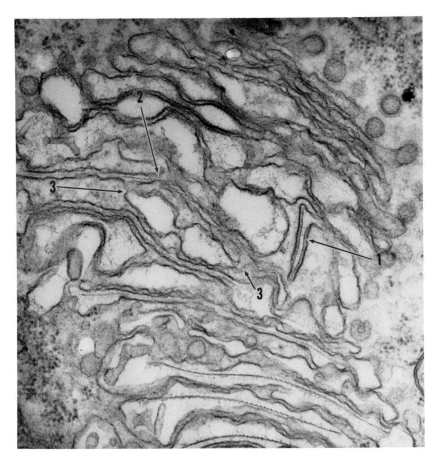

Fig. 13   Unfolding of radial network as shown
by normal post-staining techniques.   The cisterna depicted by arrow 1 shows a stage of unfolding in
which the radial network is perpendicular to the
normal scale orientation.   Arrow 2 depicts a
recently unfolded, funnel-shaped cisterna with the
two, still separate halves within it.   Possible
candidates for tubules feeding the cellulosic precursors are shown by arrows 3.   Fully synthesized
scales are shown in the lower right hand corner.
Note in the upper right hand corner the amplexus
and the budding vesicles that participate in the
forming face of the Golgi apparatus.   X 41,000.

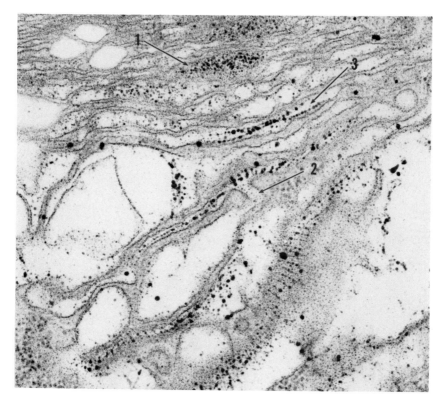

Fig. 14 Details of the presumed cellulosic
feeding tubule network (arrow 1) as revealed by the
silver technique. Radial microfibril precursor
pool shown by arrow 2. Unfolded radial stage shown
by arrow 3. X 62,000.

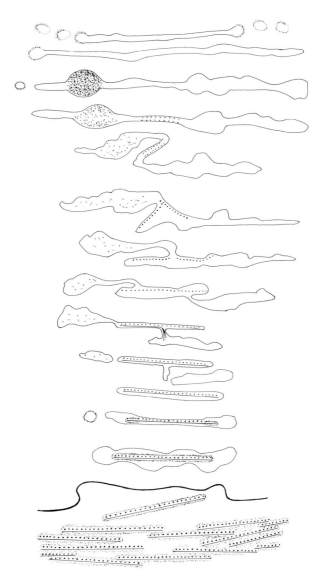

Fig. 15 Generalized scheme depicting the model of scale formation in *Pleurochrysis*. Cisternae are diagramed in median cross section to show the optimal orientation of radial and concentric microfibrils. For detailed description, see text.

242

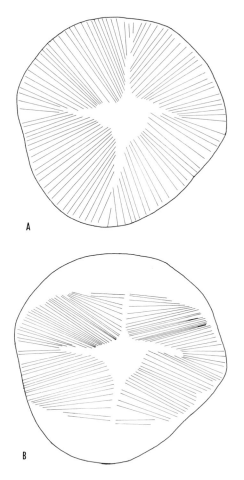

Fig. 16 Tracings of the radial microfibril network made from a negatively stained scale preparation.
16-A depicts the exact original orientation of radial
microfibrils. Note the four quadrants of microfibrils. Fig. 16-B depicts the presumed compact stage
of radial orientation immediately following unfolding. In this instance, the same microfibrils have
been re-oriented in the compact stage. (Cross sections of this compressed stage are shown in Fig. 10,
11, 12, and 14 in which the radials are folded but
still lie parallel to one another.

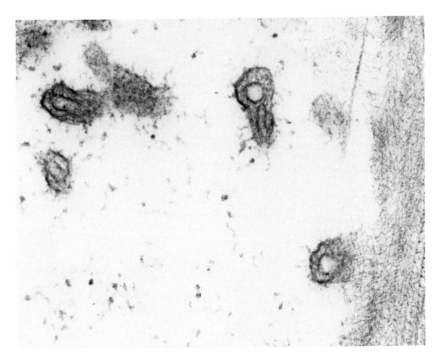

Fig. 17 Section through the finger-like projections of *Pleurochrysis* vegetative cell showing
the inner core of electron transparent material surrounded by the cytoplasmic extension of the plasma
membrane. Oblique section of cell wall shown at
right (normal post-staining). X 87,000.

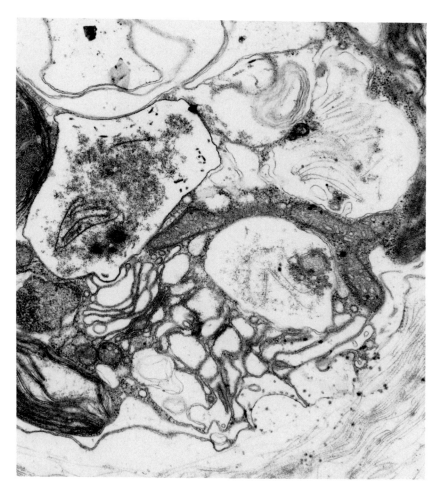

Fig. 18 Vegetative cell of *Pleurochrysis* treated with 1% colchicine for 48 hours. Note the abundance of autophagic vacuoles, the grossly disorganized cisternal network of the Golgi apparatus, and profiles of scales in the autophagic vacuoles. X 20,000.

II. Scale Chemistry

A. Carbohydrate composition of the scales

Untreated scales from scale fraction A contain radial microfibrils, amorphous material, and "concentric" microfibrils. These three components can be isolated and analyzed separately for further chemical characterization.

Thin layer chromatography of the hydrolysates from scale fraction A shows a predominance of galactose together with minor quantities of glucose and a pentose with the same R_f value as ribose, possibly fucose (4). The uronic acid content is unknown and needs to be analyzed. The amorphous and radial fibrillar material can be removed by a series of extractions. The scales have been first extracted with a sequential series of methanol and methanol/chloroform extractions (for details, 13). When the scales are treated with boiling water under magnetic stirring for two hours and centrifuged at 2,000g, the supernatant contains the dissolved amorphous material while the residue contains intact scales with radial and "concentric" microfibrils.

Further degradation occurs when the scales are treated with a 1% $NaClO_2$/1% acetic acid solution for 24 hours at 25° C under magnetic stirring. This preparation is separated and washed by repeated centrifugation and then treated with 10% NaOH solution for 24 hours under magnetic stirring at 25° C under N_2. Under these conditions the radial fibrillar material is completely removed, leaving only short rods of the "concentric" microfibrils as the purified alkali-resistant scale material corresponding to the preparation of α-cellulose from other plant sources. This material has been further analyzed chemically while the other fractions are now being analyzed in our laboratory.

Ultrastructural, cytochemical and chemical analyses of the scale preparations give the following evidence for the cellulosic nature of the alkali--resistant scale material:

1)   The *Pleurochrysis* cell wall exhibits a positive birefringence and shows a positive iodide dichroism (4).

2)   Negative staining of the purified, alkali-resistant scale material reveals microfibrils with a characteristic ribbon-like appearance and similar dimensions as known from other celluloses (3, 4).

3)   These microfibrils have a maximum flexibility of about 2% like other celluloses (4) and show typical cracking sites (breakage loci).

4)   Lateral aggregation and fasciation occurs in harsh alkali-treated preparations of scales like in other celluloses (4).

5)   This scale material is resistant to strong alkali solutions and quite resistant to strong acids like HCl, H_2SO_4 and trifluoroacetic acid (3, 4, and Herth, unpublished observations).

6)   The scale material is susceptible to acetolysis in a 1:9 mixture of concentrated H_2SO_4 and acetic acid at 100° C within less than ten minutes, thus excluding the presence of sporopollenin-like resistant material (Herth, unpublished).

7)   The alkali-resistant material is soluble in cuprammonium reagents (4).

8)   The alkali-resistant scale material is positive to zinc-chloride-iodine and stains intensively violet with this reagent (4).

9)   Total hydrolysis of the purified alkali-resistant scale material yields glucose as the predominant sugar (besides small quantities of galactose and pentose, possibly fucose). More than 95% of the sugars present in total hydrolysates is glucose as determined by glucose oxidase (4).

10) Cellobiose is the only disaccharide present after partial hydrolysis as detected by GLC of the TMS derivatives of the sugars present in the hydrolysates. The presence of laminaribiose has been excluded by comparison with a standard (13).

11) The proton resonance spectrum of the benzoylated scale material is identical with the spectrum of cotton cellulose benzoylated according to the same procedure (4).

12) The x-ray diffraction pattern of hot water--extracted scales is identical with the cellulose I diagram of the quince slime and cotton cellulose, the line broadening in scales being due to the low crystal size. After harsh alkaline treatment, the diffraction pattern of the scales corresponds to the cellulose II pattern of mercerized cellulose (13).

13) The alkali-resistant scale material yields a stable nitrate with the same solubility properties as a typical cellulose nitrate (13).

14) The degree of polymerization (=DP) as determined by fractional precipitation of the scale cellulose nitrate and viscosity measurements is 3150 with a very heterogeneous distribution of chain lengths (13).

Thus all physical, chemical and cytochemical evidence supports the cellulosic nature of scale alkali-resistant material. Of course we cannot yet exclude the presence of some linkages other than $\beta-(1\rightarrow4)$, namely $\beta-(1\rightarrow3)$ linkages. Finally, we do not know actually where the non-glucose sugars occur in the chain, but we are planning to conduct methylation analyses of the alkali resistant scale material to secure more information on this point.

B.   Amino acid composition of the alkali-
resistant scale material

Scale fraction A contains a variable amount of
N (>1%) as determined on the basis of total nitrogen content (see 4). Contamination from membraneous
fragments is very low as revealed by electron microscopy and determined by phospholipid analyses (4).
Nitrogen-containing groups in the scale polysaccharides can be excluded with certainty, since no
amino sugars have been detected. All nitrogen
present must be protein nitrogen. The nitrogen content of purified, alkali-resistant scales is even
higher corresponding to a value of 32% protein.
This value suggests a cellulose/protein ratio of
roughly 2:1 (13). The following values for the content of different amino acids recovered after total
acid hydrolysis is shown in Table 2, third column

Table 2

Amino acids (% moles) recovered from hydrolysates (6 N NCl, 36 h at 100°C) of the purified,
alkali-resistant structural polysaccharides of plants indicated at top of each column.

	Caulerpa prolifera	Lilium longiflorum	Pleurochrysis scherffelii	Cotton
	(β- (1-3)-xylan)	Pollen tube wall (β-(1-3)-glucan)	(β-(1-4)-glucan)	(β-(1-4)-glucan)
Asp X	18.4	8.2	8.4	5.6
Thr	7.0	8.2	5.6	5.6
Ser	9.6	16.5	31.9	45.0
Pro	4.7	---	traces	--
Glu X	17.8	12.4	16.8	4.2
Gly	11.9	31.1	20.7	28.1
Ala	9.4	8.2	7.8	11.2
Val	8.7	0.4	1.3	--
Ile	2.0	6.2	2.2	--
Leu	5.5	8.2	3.9	--
Try	3.3	--	traces	--
Phe	--	--	traces	--
Met	1.6	--	--	--
Arg	+	+	+	+
Lys	traces	traces	traces	traces
His	traces	traces	traces	traces

249

(*Pleurochrysis*). In *Pleurochrysis*, about 38% of the
amino acids contain hydroxyl groups with serine as
the major amino acid, and about 25% of the amino
acids contain carboxyl groups, with asparagine as
major amino acid. The original serine content might
have been even higher since serine is well known to
be converted partially to glycine under the alkaline
conditions used for purification of the material.
A very high content of ammonia in the hydrolysates
and the neutral behavior of the scales in a titra-
tion assay (Herth, unpublished observations) speak
for asparagine and glutamine and not the corresponding acids, to be originally present in the native
scale fraction.

The microfibrils of alkali-resistant scale
material do not show any loss of N content when
stirred with 8 M urea for 24 hours at 25° C or with
5 M guanidinium chloride. This is indicative of a
true covalent linkage of peptide to the polysaccharide moieties. As far as the type of linkage is
concerned, we can only suggest that the linkage is
via an ether bond between a sugar hydroxyl group
and the hydroxyl group of an amino acid or from a
sugar carbon to the nitrogen atom of asparagine or
glutamine. We have not yet been able to detect
linked sugar-amino acid pieces with thin layer
chromatography from hydrolysates of the alkali-resistant scale material. In addition, we have been
unsuccessful obtaining such moieties by enzymatic
degradation, since no enzyme yet tested (compare
Herth et. al., 13 and Green and Jennings, 11) including cellulase and snail digestive enzyme has
degraded the alkali-resistant scale microfibrils*.
Thus is is still an open question whether glucose
or galactose or both sugars are linked to an amino

---

*Cellulase now has been shown to degrade concen-
trics (Herth, et al., in preparation).

acid. Future experiments include a β-elimination assay under alkaline conditions to determine the involvement of serine. In addition, we are planning to analyze the acid hydrolysates by gas chromatography to determine whether we can detect sugar--amino acid linked moieties.

III. Peptides Linked to Other Structural Polysaccharides

To see whether this very intimate association of peptide moieties with the structural polysaccharide in *Pleurochrysis* scales is a unique occurrence or whether it is a more general phenomenon, the investigations have been extended to include structural polysaccharides from other organisms. These polysaccharides have been purified according to the same extraction procedures described for *Pleurochrysis*. These structural polysaccharides have been identified by hydrolysis, x-ray diffraction, and NMR (Herth et. al., in preparation). The alkali-resistant structural polysaccharides thus far isolated include (see also Table 2):

β-(1→3)-xylan from *Caulerpa prolifera*
β-(1→4)-mannan from *Acetabularia mediterranea*
β-(1→3)-glucan from pollen tube walls of
         Lilium longiflorum
β-(1→4)-glucan (cellulose) from onion root tips
β-(1→4)-glucan from cotton (Eucellulose standard)

Negative staining in all cases reveals a microfibrillar structural polysaccharide to be present in the final, alkali-resistant fraction (Herth, et. al., in preparation). No amino sugars have been detected in these hydrolysates, but amino acids were detected by thin layer chromatography, gas chromatography and automatic amino acid analysis.

251

Table 2 summarizes the amino acids recovered from acid hydrolysates of alkali-resistant materials from *Caulerpa prolifera*, lily pollen tube wall, and cotton in comparison with the *Pleurochrysis* scale material (*Acetabularia* mannan and onion root glucan are just being analyzed now). In all cases, we see the predominance of acidic amino acids (especially with *Caulerpa* xylan) as well as hydroxy amino acids. There is an obvious similarity of the amino acid composition of the peptides linked to *Pleurochrysis* alkali-resistant β-(1$\rightarrow$4)-glucan and the peptides linked to cotton cellulose.

The alkali-resistant pollen tube wall material can be completely dissolved in N-ethyl-pyridinium--chloride (Herth, unpublished), a potent solvent medium for cellulose and other structural polysaccharides (14). Furthermore, this material can be re-precipitated with methanol. The reprecipitated material still contains the peptide moieties, thus providing more evidence for covalent linkage of the peptides to the structural polysaccharide. (The same type of experiment will be used to test the other structural polysaccharides mentioned above). These preliminary results suggest that the peptide moieties intimately associated with the structural polysaccharide in *Pleurochrysis* are not a unique case among the plant kingdom. Thus, is is a good model for comparison with all other cases of structural polysaccharides, and we have the feeling that no really pure structural polysaccharide exists.

Discussion

While it is generally accepted that the Golgi apparatus functions as an assembly line for complex macromolecules such as mucopolysaccharides, hemicelluloses, glycoproteins, proteoglycans, and glycolipids (23, 25, 34), the role of this organelle in cellulose biosynthesis has been excluded. The classical view of cellulose biosynthesis, based on

Colvin's (8) work with bacterial cellulose, holds
that this structural polysaccharide is polymerized
and crystallized extracellularly or on the surface
of the plasma membrane via diffusion of non-membrane
cytoplasmic pools of nucleoside diphosphate sugar
compounds. On the contrary, recent evidence has
suggested a role of the Golgi apparatus in the
biosynthesis of cellulose. Evidence has come from two
different approaches and with two different cellular
systems: Ray (30) has demonstrated a β-(1→4)-glucan
synthetase to be associated with the Golgi fraction
of pea seedlings. Brown and co-workers (3, 4) have
found a cellulosic moiety in Golgi-produced scales
of a marine alga. While it is now well-known that
the Golgi apparatus is involved in scale formation
(17, 20) the extended observations reported here
permit one to make a more precise working hypothesis
about the role of this organelle in polysaccharide
biosynthesis. It is possible with *Pleurochrysis* to
differentiate the specific sites of polymerization
and assembly of the morphological scale sub-components
and to characterize them chemically.

At present, we think that it has been clearly
demonstrated that the scales have a cellulosic
sub-unit structure in the form of the so-called,
"concentric" microfibrils, The microfibrils are
not pure cellulose but intimately associated with
peptide moieties. In addition, they are not synthesized alone but deposited on other structured polysaccharide moieties known as radial microfibrils,
and later coated by a slimy, presumably pectic moiety
known as the amorphous material that also cross-
-links the scales to produce a coherent cell wall.
The different assembly steps of these three sub-
-components can be followed by electron microscopy,
and the observations indicate that a very complex
machinery is necessary to synthesize the scale network by the stepwise addition of new components.
The entire scale assembly process requires only
two minutes (2), a relatively short time requirement

253

demonstrating the great efficiency of the Golgi machinery in this instance.

Generalizing with this model we can assume that in the Golgi apparatus of higher plants, structural polysaccharides are not synthesized alone but in association with peptides and acidic polysaccharides. The structural polysaccharides could thus be prevented from immediate crystallization and after secretion, could co-crystallize ultimately producing the crystalline microfibrils of the mature cell wall. Lysosome-like dense vesicles could secrete enzymes for removing the coat of acidic polysaccharides, resulting in spontaneous lateral aggregation of the "naked" structural polysaccharide polymers. The peptide moieties associated with the structural polysaccharides are thought to have two possible roles:

a) certain peptides could be involved in cross-
    -linking of the structural polysaccharides,
    thus stabilizing special arrangements of the
    microfibrils (16). This is exemplified in the
    Pleurochrysis model by the lateral aggregation
    of six to ten cellulosic microfibrils into a
    definitive band, and by the spiral arrangement
    of these bands to form the scale skeleton.

b) Other peptides could function as initiators or
    recognition sites for specific sugar additions,
    resulting in a direct genetic control of the
    type of polysaccharide being synthesized. This
    could be analogous to the postulated role of
    specific sequences of amino acids in animal
    glycoprotein peptides adjacent to the amino
    acid-sugar linkages (21, 32).

We believe that the biosynthesis of animal glycoproteins (see 33), plant glycoproteins (7, 31) and structural polysaccharides covalently-linked to peptide moieties (13) follow the same general pathway even though the ratio of carbohydrate/protein content may be much higher in plants.

Another interesting question concerns the function necessary for obtaining a distinct degree of polymerization of the structural polysaccharides (compare the template hypothesis of Marx-Fignini, 22). From the *Pleurochrysis* model we can speculate about some possible control mechanisms: (a) cisternal sac size; (b) membrane-attached synthetase activity and/or availability; (c) number of peptides functioning as glucan initiators; (d) precursor pool size; (e) length and number of radial microfibrils formed to function as the first ultra-structurally visible template for the deposition of the spiral bands of microfibrils; and (f) the time-limitation imposed for any given step of the assembly line process. Experiments are now in progress to clarify the proposed hypothesis of control mechanisms in scale formation. From these data we hope to produce a model which will be directly applicable to the eventual understanding of higher plant cell wall biosynthesis.

Acknowledgments

The authors would like to thank the Laboratories for Reproductive Biology, University of North Carolina, Chapel Hill, the National Science Foundation (GB 23047 to R.M.B.), Deutsche Forschungsgemeinschaft (to W.H. and W.W.F.), and the NDEA (to D.R.) for generous financial assistance.

W.H. wishes to thank Professor E. Husemann for many discussions and support of this work. For the amino acid analyses, we thank Prof. Witt from the Childrens Hospital of the University of Freiburg, Germany. We thank Professors H. Stanley Bennett, Peter Sitte, Dr. Heinz Falk and Mr. Richard Santos for helpful discussions.

References

1. J. R. Barnett and R. D. Preston, *Ann. Bot. 34*, 1011 (1970).
2. R. M. Brown, *J. Cell Biol. 41*, 109 (1969).
3. R. M. Brown, W. W. Franke, H. Kleinig, H. Falk, and P. Sitte, *Science 166*, 894 (1969).
4. R. M. Brown, W. W. Franke, H. Kleinig, H. Falk, and P. Sitte, *J. Cell Biol. 45*, 246 (1970).
5. R. M. Brown, *Encyclopedia Cinematographia*, Institut für den Wissenschaftlichen Film, Göttingen (1970).
6. R. M. Brown and W. W. Franke, *Planta 96*, 354 (1971).
7. M. J. Crispeels, *Biochem. Biophys. Res. Commun. 39*, 732 (1970).
8. J. R. Colvin in M. H. Zimmermann ed. *The Formation of Wood in Forest Trees*, New York, Academic Press, (1964), p. 189.
9. W. W. Franke, *Planta 90*, 370 (1970).
10. W. W. Franke and R. M. Brown, *Arch. Microbiol. 77*, 12 (1971).
11. J. C. Green and D. H. Jennings, *J. Exp. Bot. 18*, 359 (1967).
12. W. Herth, W. W. Franke, and W. J. Vanderwoude, *Naturwissenschaften 59*, 38 (1972).
13. W. Herth, W. W. Franke, J. Stadler, H. Bittiger, G. Keilich, and R. M. Brown, *Planta 105*, 79 (1972).
14. E. Husemann and E. Siefert, *Die Makromolekulare Chemie 128*, 288 (1969).
15. H. D. Isenberg, S. D. Douglas, C. S. Lavine, S. S. Spicer, and H. Weissfellner, *Ann. N. Y. Acad. Sci. 136*, 155 (1966).
16. D. T. Lamport, *Ann. Rev. Plant Physiol. 21*, 235 (1970).
17. I. Manton, *J. Cell Sci. 2*, 265 and 411 (1967).
18. I. Manton and G. F. Leedale, *J. mar. biol. assn. U.K. 49*, 1 (1969).
19. I. Manton and M. Parke, *J. mar biol. assn. U.K. 45*, 743 (1965).

20. I. Manton and L. S. Peterfi, *Proc. Roy. Soc.* (London) *B, 172*, 1 (1969).
21. R. D. Marshall and A. Neuberger, *Advanc. Carbohydr. Chem. Biochem. 25*, 407 (1970).
22. M. Marx-Figini, *Biochim. Biophys. Acta 177*, 27 (1969).
23. D. H. Northcote, *Endeavour 30*, No. 109, 26 (1971).
24. D. H. Northcote, *Essays in Biochemistry, 5*, 90 (1969).
25. D. H. Northcote and J. D. Pickett-Heaps, *Biochem. J. 98*, 159 (1966).
26. D. E. Outka and D. C. Williams, *J. Protozool. 18*, (2), 285 (1971).
27. B. Parker and A. G. Diboll, *Phycologia 6*, 37 (1966).
28. J. D. Pickett-Heaps, *J. Histochem. Cytochem. 15*, 442 (1967).
29. R. N. Pienaar, *J. Phycol.* 5, 321 (1969).
30. P. Ray, T. L. Shininger and M. M. Ray, *Proc. Nat. Acad. Sci. U.S. 64*, 605 (1969).
31. J. Savada and M. J. Chrispeels, see chapter 9, this book.
32. R. G. Spiro, *Ann. Rev. Biochem. 39*, 599 (1970).
33. A. Weinstock and C. P. Leblond, *J. Cell. Biol. 51*, 26 (1971).
34. W. G. Whaley, M. Dauwalder and J. E. Kephart, *Science 175*, 596 (1972).

STUDIES ON THE BIOSYNTHESIS OF CELLULOSE

A. D. Elbein and W. T. Forsee

Department of Biochemistry
The University of Texas Medical School
San Antonio, Texas   78229

Abstract

A particulate enzyme was isolated from mung bean seedlings which catalyzed the incorporation of glucose from GDP-^{14}C-glucose into an alkaline-insoluble polymer. On the basis of isolation of ^{14}C-cellobiose and ^{14}C-cellotriose by partial acid hydrolysis or acetolysis and on the production of ^{14}C-erythritol by Smith degradation, the alkaline-insoluble polymer appears to be a β-(1 →4)-glucan (cellulose). UDP--^{14}C-glucose is also incorporated into polysaccharide by this particulate enzyme but most of this product is soluble in alkali, although small amounts of radioactivity are found in alkali-insoluble material. Both the alkali-soluble and alkali-insoluble products formed from UDP-^{14}glucose have been characterized as β-(1→ 3)-glucans. The incorporation of radioactivity from GDP-^{14}C-glucose into insoluble products is stimulated 3 to 4 fold by the addition of unlabeled GDP-mannose. Radioactivity from GDP-^{14}C-mannose is also incorporated into alkaline-insoluble material. The product is formed from GDP-^{14}C-mannose and the product formed from GDP-^{14}C-glucose (plus GDP-mannose) have been characterized as β-(1→ 4)-glucomannans (with the possibility that some β-(1→ 4)-mannan is also formed). The particulate enzyme was subjected to sucrose density gradient centrifugation and all of these enzymatic activities were located in the same

fraction which had a density equal to 25% sucrose.
These particles while not homogeneous had the appearance of membranous vesicles in the electron microscope.

Two glycolipids were synthesized by extracts of
cotton fibers, one from GDP-^{14}C-mannose and the
other from UDP-^{14}C-glucose. Both of these lipids
are acidic and are retained on DEAE-cellulose and
both have Rf's on thin layer chromotography which
are similar to Rf's reported for bacterial lipids
which are intermediates in polysaccharide synthesis
(mannosyl-phosphoryl-polyprenols). The sugar moieties
of both lipids are very lable to acid hydrolysis
(0.003 -0.01 N HCl, 100° for 1-5 min) and the mannolipid has a phosphate to mannose ratio of about 1:1.
Finally the mannose portion of the mannolipid can
be transferred to GDP to form GDP-mannose and the
glucose portion of the glucolipid to UDP to form
UDP-glucose. Similar types of glycolipids were
isolated from intact cotton bolls. An enzyme, free
of endogenous lipids was isolated, and this enzyme
required the addition of an active lipid acceptor
in order to incorporate radioactivity from GDP-[14]C-
-mannose into chloroform-methanol.

Symbols

ADP-^{14}C-glu, etc. = ADP-^{14}C-glucose, etc.
GDP-^{14}C-man   = GDP-^{14}C-mannose
GDP-man   = GDP-mannose
GDP-glu   = GDP-glucose
UDP-glu   = UDP-glucose

Introduction

Plant cells contain a true cell wall which is
composed largely of cellulose but also contains
various other polysaccharides, as well as lignins,
tannins and possibly small amounts of proteins and
lipids. In cells of soft tissue only a primary cell

wall is laid down, whereas in woody and fibrous tissues a secondary wall is formed after and interior to the primary wall (1, 2).

Cell walls normally contain cellulosic fibrils that are embedded in a matrix of hemicellulosic polysaccharides and pectins. In the primary cell wall, cellulose is supposed to have a random and lower molecular weight than in the secondary wall (3). The diagram in Figure 1 is a schematic representation of a plant cell-wall taken from Dadswell

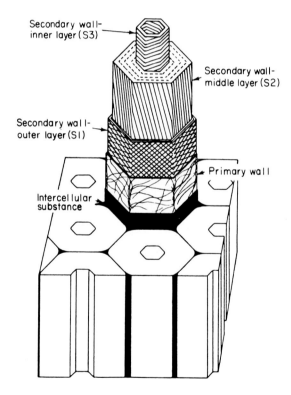

Fig. 1 An idealized representation of a typical plant cell wall showing the primary and secondary cell wall layers (from Dadswell and Wardrop (1960)).

and Wardrop (4). This diagram shows the amorphous intercellular or middle lamella containing mostly lignin; the primary wall with a loose network of microfibrils embedded in a matrix composed of hemicellulose and pectin, and the various layers of a secondary wall (S1-S3). The S1 layer is usually very thin with fibrils arranged in a cross-hatched pattern; S2 constitutes the bulk of the secondary wall and is composed of parallel fibrils and the S3 layer which also contains parallel fibers; S3 also contains parallel fibrils but in a different orientation from S2.

The chemical composition of cellulose isolated from plant cell walls has been shown to a linear polymer of D-glucose units linked in β-(1→ 4) glycosidic bonds. Thus, the fibrils in these various layers represent regions of cellulose of about 70% crystallinity. However, it should be pointed out that after removal of the amorphous materials from plant cell walls, the residue which remains is not necessarily "pure cellulose" as defined by the organic chemist, but still contains small amounts of other sugars such as xylose, mannose, galactose, etc. (5). These other sugars are regarded as an integral part of the microfibrils and are not thought to be contaminants due to hemicelluloses (5). In fact, it has been suggested that these other sugars may represent or account for the non-crystalline (paracrystalline) regions of the fibrils (6). Thus, the crystalline regions would be composed of D-glucose attached by β-(1→ 4) linkages. These crystallites would be attached to each other through amorphous areas made up of sugars other than glucose (i.e., mannose, uronic acids, etc.).

Biosynthesis of Cellulose

In 1964, Elbein, Barber and Hassid (7) began an investigation on the biosynthesis of cellulose in mung bean seedlings (*Phaseolus aureus*). In these experiments, various ^{14}C-glucose nucleotides

(ADP-^{14}C-glu, CDP-^{14}C-glu, GDP-^{14}C-glu,TDP-^{14}C-glu, UDP-^{14}C-glu) were prepared chemically by the mor-pholidate procedure (8) and each of these ^{14}C-nuc-leotides was then incubated with a particulate enzyme fraction obtained from mung bean shoots. After incubation, the insoluble material was isolated by centrifugation, extracted several times with hot alkali to remove protein and hemicelluloses, washed with water, and then counted to determine its radio-active content. As shown in Table I, radioactivity was found in the alkaline-insoluble residue when GDP-^{14}C-glu was used as substrate, but the other sugar nucleotides and sugars were inactive as glu-cosyl donors. Although the amount of radioactivity incorporated into the alklaine-insoluble material from GDP-^{14}C-glu was quite significant, the actual amount of sugar incorporated (in terms of μmoles of sugar) was very low since the sugar nucleotides had very high specific activities.

The insoluble product formed from GDP-^{14}C-glu was characterized as a β-(1→ 4) linked glucan (cellulose) on the basis of the data shown in Table II (9). The product was insoluble in water and alkali but could be solubilized in syrupy phosphoric acid and cuprammonium hydroxide (cuoxan) and was precipitated when these solvents were diluted with water. Partial acid hydrolysis of the ^{14}C-product with fuming HCl gave rise to a series of oligosaccharides which had the same mobility as those produced by partial acid hydrolysis of authentic cellulose (Figure 2). The ^{14}C-disaccharide had the same mobility as cellobiose in several different solvent systems and was co-cry-stallized with authentic cellobiose to constant spe-cific activity. Both the disaccharide and the tri-saccharide were hydrolyzed to glucose by β-gluco-sidase (almond emulsin), and the trisaccharide gave rise to a ^{14}C-tetrose and ^{14}C-cellobiose upon oxida-tion with lead tetraacetate. Finally, acetolysis of the product followed by deacetylation and chroma-tography gave rise to a ^{14}C-disaccharide with the

TABLE I

Substrate Specificity For Cellulose Synthesis[a]

C^{14}-glucosyl donor	Product, c.p.m.
Guanosine diphosphate D-glucose	1165
Uridine diphosphate D-glucose	0
Thymidine diphosphate D-glucose	4
Adenosine diphosphate D-glucose	0
Cytidine diphosphate D-glucose	9
D-glucose	0
D-glucose 1-phosphate	0

[a]The complete incubation mixture contained the following in a
total volume of 0.2 ml.: 0.5 μmole of 0.1 M̲ Tris buffer (pH 7.6),
2 X 10^{-4} μmole (12,000 c.p.m.) of the indicated glucosyl donor
(106 μc/μmole), and 0.1 ml. of enzyme representing about 2 g.
wet weight of plant material (about 1.5 mg. of protein and 0.1 mg.
of cellulose). The mixtures were incubated at 37^{0} for 30 min.
At the end of this time, 1 ml. of water was added, and the reactions
were stopped by heating for 3 min. in a boiling water bath. One mg.
of cellulose was added, and the precipitate was collected by
centrifugation. The precipitate was again extracted with hot
water and was then heated two times (100^{0} for 5 min.) with 1 ml.
portions of 2% NaOH. The precipitate was washed with 1ml. of
water, suspended in 0.2 ml. of water, and plated on planchets.
Radioactivity in the samples was determined wtih a mica-window
Geiger-Muller counter and conventional scaler.

PLANT CELL WALL POLYSACCHARIDES

TABLE II

Characterization of Enzymatically Synthesized Cellulose

1. Radioactive product was insoluble in most solvents but
   could be dissolved in syrupy phosphoric acid and cuoxan.

2. Partial acid hydrolysis and chromatography of oligosaccharides.

3. Cochromatography of ^{14}C-disaccharide with cellobiose.

4. Crystallization of disaccharide to constant specific activity
   with authentic cellobiose.

5. Hydrolysis of di and trisaccharide with β-glucosidase to
   glucose.

6. Production of erythrose and cellobiose upon oxidation of trisaccharide with lead tetracetate.

7. Acetolysis of the polysaccharide and subsequent production of
   cellobiose.

265

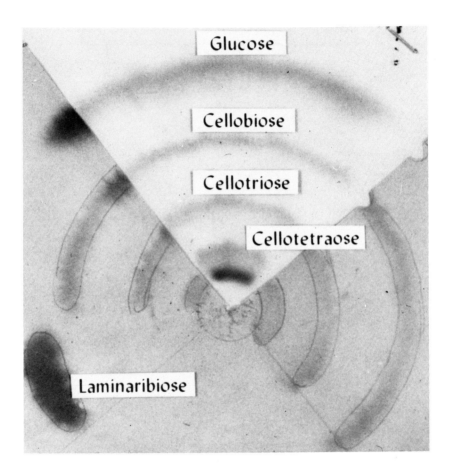

Fig. 2 Radial chromatogram of a partial acid hydrolysate of enzymatically synthesized cellulose. Radioactive compounds were detected by exposure of the chromatogram to X-ray film and authentic oligosaccharides were visualized with p-anisidine phosphate.

mobility of cellobiose. Treatment of cellulose under similar conditions gave cellobiose as the major product. Thus, all of these criteria indicate that the ^{14}C-glucose which is incorporated into the alkali-insoluble product is linked by β-(1$\rightarrow$ 4) glycosidic bonds presumably to an acceptor cellulose molecule. Based on the lead tetraacetate studies, at least two radioactive glucose molecules are added to the end of these chains. As shown in Table III, enzyme fractions obtained from a variety of plants were active in the incorporation of GDP-^{14}C-glu into alkaline--insoluble polysaccharides.

TABLE III

Incorporation of GDP-^{14}C-Glu into Alkaline-Insoluble
Polysaccharide by Various Plant Seedlings

Seedling	CPM in Product/mg. Protein
Mung Bean	2362
String Bean	1601
Squash	1067
Peas	741
Corn	521

Brummond and Gibbons (10) reported that a part-iculate enzyme preparation from *Lupinus albus* catalyzed the incorporation of radioactivity from UDP-^{14}C-glu into heterogeneous product which ranged in size from water-insoluble to alkali-insoluble. The product was identified as cellulose by acetolysis and isolation of the cellobiose octaacetate. However, the conditions of acetolysis used by these workers apparently cleaves β-(1$\rightarrow$ 3)-glycosidic bonds since these are less stable than β-(1$\rightarrow$ 4) bonds. These workers also found that GDP-glu was incorporated into a product with the characteristics of cellulose but at a lower rate than was UDP-glu. Ordin and Hall (11) found that UDP-glu was a more effective sub-

strate for polysaccharide synthesis than was GDP-glu
with a particulate enzyme from *Avena sativa*. However,
enzymatic hydrolysis of the UDP-glu product produced
a number of unidentified products whereas the GDP-
glu product gave rise almost exclusively to cellobiose. Batra and Hassid (12) and Flowers et al (13)
were only able to demonstrate the formation of
β-(1→ 3) linkages from UDP-glu whereas GDP-glu
gave a β-(1→ 4) glucan. Lenoel and Axelos (14)
reported that both β-(1→ 3) and β-(1→ 4) linkages
may be obtained from UDP-glu with the same enzyme
preparation from wheat roots depending on the
conditions used in these experiments and especially
on the concentration of the substrate. It thus appears from these various experiments that GDP-glu
is the substrate for cellulose-like glucans whereas
UDP-glu may give rise to both β-(1→ 3) and β-(1→ 4)
linkages possibly in the same polymer. In general,
the UDP-glu:polysaccharide transferase appears to be
much more active in these various tissues than is
the GDP-glu:polysaccharide transferase. Feingold,
et. al., (15) previously showed that UDP-glu was a
substrate for the formation of callose, but it seems
probable that it also is a substrate for a mixed
polysaccharide.

Because of these various discrepancies, Chambers
and Elbein (16) reexamined the polysaccharide synthesizing system in mung bean seedlings and compared
the products formed from both UDP-^{14}C-glu and GDP-
[14]C-glu. As shown in Figure 3, most of the radioactivity incorporated into polysaccharide from GDP-
glu was insoluble in alkali whereas most of the
radioactivity incorporated from UDP-glu into water
insoluble polysaccharides was solubilized by extraction with hot alkali. This is shown more quantitatively in Table IV which presents data from a
large-scale incubation. In this case, less than 5%
of the GDP-glu product was removed by alkali extraction whereas more than 90% of the UDP-glu product was solubilized by this treatment. However,

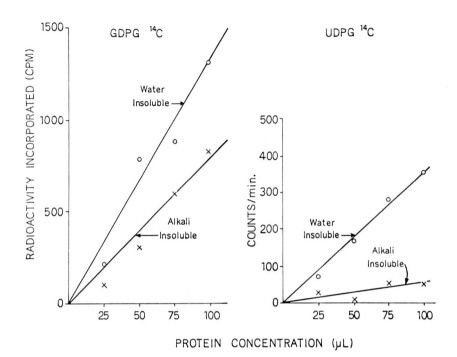

Fig. 3 Effect of protein concentration on the incorporation of radioactivity from UDP-^{14}C-glu and GDP-^{14}C into water-insoluble and alkali-insoluble polysaccharides. Reactions were as described in the text except that the amount of the particulate enzyme fraction varied. Incorporation into the water-insoluble (0 - 0) and alkali-insoluble (X - X) fractions were measured as indicated in the text. Specific activities of sugar nucleotides were as follows: UDP-^{14}C-glu, 125,000 cpm/μmole; GDP-^{14}C-glu, 2 X 10^6 cpm/μmole. Therefore, an incorporation of 500 cpm represents about 4 X 10^{-3} μmole of UDP-^{14}C-glu and about 2 X 10^{-4} μmole of GDP-^{14}C-glu.

TABLE IV

Incorporation of Radioactivity from UDP-^{14}C-Glu and
GDP-^{14}C-glu into Various Polysaccharide Fractions

Product	Radioactivity Incorporated from GDP-^{14}C-glu	UDP-^{14}C-glu
Alkali-Insoluble Fraction	274,000	10,000
Alkali-Soluble, water-insoluble fraction (Hemicellulose A)	12,000	100,000

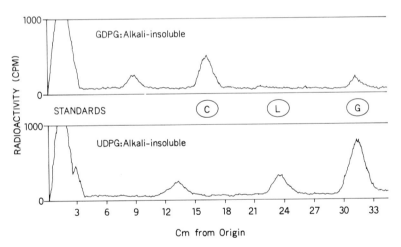

Fig. 4 Radioactive tracings of the products formed by partial acid hydrolysis. The oligosaccharides
were separated by paper chromatography in n-propanol:
ethyl acetate:water (7:1:2). Upper tracing:alkali-
insoluble polysaccharide formed in the presence of
UDP-^{14}G-glu; Lower tracing:alkali-insoluble polysaccharides formed from UDP-^{14}C-glu. Standards are
C = cellobiose, L = laminaribiose, G = glucose.

significant amounts of radioactivity remained in the alkaline-insoluble material from UDP-glu suggesting that solubility alone is not a good indication of the presence of cellulose. The ^{14}C-products (both alkaline-soluble and alkaline-insoluble) formed from both GDP-glu and UDP-glu were subjected to partial acid hydrolysis and the oligosaccharides were isolated by paper chromatography. Figure 4 shows a radioactive tracing of these hydrolysates. It can be seen that the product formed from GDP-glu gave radioactive peaks corresponding to cellobiose and cellotriose whereas the UDP-glu product gave radioactive peaks corresponding to the laminarin oligosaccharides. Similar results were obtained when the polysaccharides were degraded by the acetolysis procedure. The larger oligosaccharides obtained from acetolysis of both polysaccharides as well as the insoluble polysaccharides themselves were subjected to periodate oxidation and were then reduced with NaBH$_4$ (Smith degradation). After acid hydrolysis, the alcohols and sugars were separated by paper chromatography. Figure 5 shows a tracing of the radioactive products formed from both polymers. Oxidation of the GDP-glu product gave rise to ^{14}C-erythritol indicating the presence of 1→ 4 linkages whereas oxidation of the UDP-glu product gave radioactive peaks corresponding to glucose and glycerol, but no radioactive erythritol. The fact that most of the radioactivity in this latter case was in glucose would suggest either a 1→ 3 linked polymer or a branched structure. The erythritol formed from the GDP-glu product was further identified by gas liquid chromatography. Therefore under the conditions used in these studies, GDP-glu appears to be a precursor for β-(1→ 4) glycosidic bonds whereas UDP-glu is the substrate for a β-(1→ 3) linked polysaccharide.

We have also examined the mechanism of cellulose synthesis in maturing cotton fibers. Franz and Meier (17) reported that UDP-^{14}C-glu was incorporated

271

into cellulose by whole cotton fibers, but our results are not in agreement with these. We compared
the products formed from both UDP-^{14}C-glu and GDP-
[14]C-glu and obtained essentially the same results as
described above for mung beans. GDP-glu was incorporated into an alkali-insoluble polymer which was
characterized as a β-(1→4) glucan by isolation and
characterization of hydrolysis products (oligosaccharides) and by the production of erythritol by
Smith degradation. On the other hand, most of the
radioactivity from UDP-glu was found in alkali-
soluble products, but all of the UDP-glu incorporated
(including that small amount in alkaline-insoluble
material) was found in 1→3 linkages with possibly
small amounts of 1→4 linkages. The peak of activity
of the GDP-glu:β-(1→4)-glucan transferase occurred
in 10-day old bolls and then declined so that by 20
days there was very little enzymatic activity remaining in these tissues (Figure 6). This period
of time corresponds to the time when the cotton
fiber is elongating and producing primary cell wall.
After about 20 days of age (of cotton bolls), the
fiber stops elongating and begins to lay down the
secondary wall. Neither the peak activity of
incorporation with either GDP-glu or UDP-glu appeared
to coincide with this latter phase of development.
However, it seems possible that one sugar nucleotide
such as GDP-glu is the precursor to the cellulose
in the primary cell wall, and that another nucleotide
is the substrate for secondary cell-wall cellulose.
This could explain some of the disparities mentioned
above. However, we have not been able to demonstrate
the incorporation of radioactivity from other glucose
nucleotides into alkaline-insoluble material using
cell-free extracts from older cotton fibers.

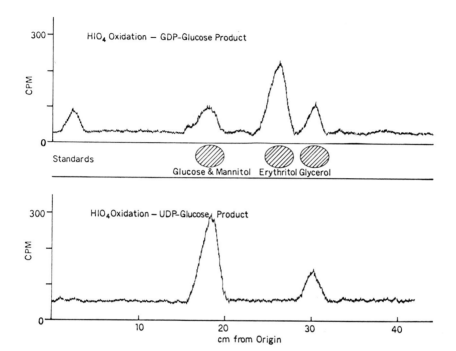

Fig. 5 Radioactive tracing showing the products of periodate oxidation (Smith degradation) of the ^{14}C-polysaccharides. Following reduction and hydrolysis, degradation products were chromatographed on paper in Solvent III. Upper tracing:[14]C-products from Smith degradation of the polymer formed from GDP-[14]-C-glu; Lower tracing:[14]C-products from the polymer synthesized from UDP-^{14}C-glu.

273

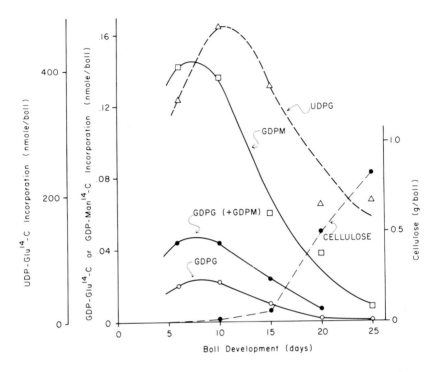

Fig. 6 Developmental profile of polysaccharide synthesizing enzymes in cotton fibers.

PLANT CELL WALL POLYSACCHARIDES

Stimulation of GDP-glu
Incorporation by GDP-man

During the course of studies on the biosynthesis
of cellulose, it was observed that the addition of
unlabeled GDP-man markedly stimulated the incorporation of radioactivity from GDP-^{14}C-glu into
alkali-insoluble material (Figure 7). Addition of
GDP-man caused a three to four fold increase in the
incorporation of radioactivity from GDP-^{14}C-glu. It
should also be noted that radioactivity from GDP-^{14}C-
man was incorporated to a much greater extent than
was GDP-^{14}C-glu even in the presence of GDP-man. The
alkali-insoluble product formed from GDP-^{14}C-man was
characterized as a β-(1→ 4)-linked glucomannan (and
possibly a β-(1→ 4)-mannan) on the basis of chemical
characterization of the oligosaccharides released by
various hydrolytic techniques. For example, Figure
8 shows radioactive scans of the oligosaccharides
released from this polymer by enzymatic hydrolysis
(using a β-mannanase) or partial acid hydrolysis or
acetolysis. Each of these methods gave the same
types of oligosaccharides but in different amounts.
Thus, two disaccharides were obtained, one of which
had equal amounts of glucose and mannose and the
other which only contained mannose. The larger
oligosaccharides were subjected to Smith degradation
(periodate oxidation and NaBH$_4$ reduction) and the
alcohols released by hydrolysis were isolated by
paper chromatography. Two major radioactive peaks
were observed which corresponded to erythritol and
glycerol. The occurence of erythritol confirms the
presence of 1→ 4 glycosidic linkages. It therefore appears that the mannose is incorporated into a
β-(1→ 4)-linked glucomannan and possibly into a
β-(1→ 4)-mannan. The possibility that a mannan is
also formed is based on the isolation of a disaccharide containing only mannose. The insoluble polymer formed from GDP-^{14}C-glu and GDP-man also gave
rise to a disaccharide of glucose and mannose by enzymatic hydrolysis and yielded ^{14}C-erythritol by

275

Smith degradation also suggesting a β-(1→ 4) glucomannan (18).

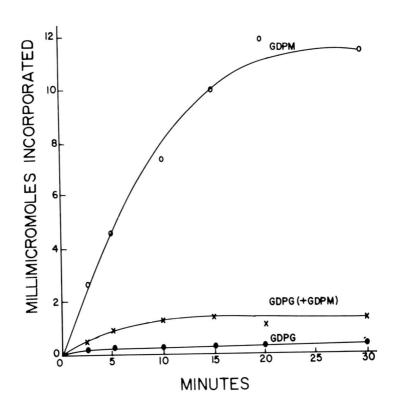

Fig. 7 The effect of time on the incorporation
of radioactivity from GDP-^{14}C-man, GDP-^{14}C-glu and
GDP-^{14}C-glu plus GDP-man into the alkali insoluble
polysaccharides.

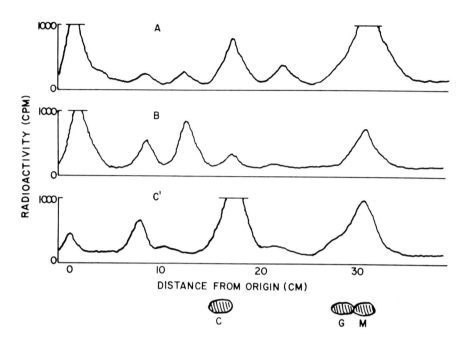

Fig. 8. Radioactive tracing of the soluble oliogosaccharides released from the polymer. The glucomannan, synthesized from GDP-^{14}C-man, was treated as follows: Tracing A, enzymatic hydrolysis; Tracing B, partial acid hydrolysis; Tracing C', acetolysis. The soluble oligosaccharides were separated by paper chromatography on Whatman No. 33 paper in Solvent III. The polymer synthesized from GDP-^{14}C-glu and GDP-man was also subjected to enzymatic hydrolysis and a somewhat similar scan was obtained except that considerably less radioactivity was in the various oligosaccharides. Standard compounds shown under the line are: C, cellobiose; G, glucose; M, mannose.

Experiments were done in an attempt to determine the sequence of addition of sugars and the effect of mannose addition on glucose incorporation. As shown in Figure 9, when the particulate enzyme was incubated with GDP-^{14}C-glu alone, there was a rapid incorporation of radioactivity into alkaline-insoluble products which stopped after 5 minutes. The reason for the termination of this reaction appears to be that the acceptor molecules are all saturated. Thus the addition of GDP-^{14}C-glu in a second incubation did not result in any additional incorporation of radioactivity. However, when GDP-man was added 5 min before washing out the first incubation mixture, the addition of GDP-^{14}C-glu in the second incubation produced a second burst in the incorporation of radioactivity suggesting that the incorporation of mannose provided new acceptor sites for glucose. GDP-man did not have to be present at the same time as GDP-glu in order to stimulate the incorporation. Thus, the particles could be incubated with GDP-man, then isolated by centrifugation and washed to remove the residual GDP-man and then incubated with GDP-^{14}C-glu. Under these conditions, the incorporation of glucose was still stimulated.

Although, it appears from these experiments that GDP-man stimulates the incorporation of glucose from GDP-^{14}C-glu into a glucomannan, it may be that the mannose and glucose are really part of the cellulose molecule as it exists in nature. That is, the mannose may represent those amorphous regions of the cellulose chain and synthesis may actually be initiated at these points. The problem with "in vitro" biosynthetic experiments is that the only products which can easily be examined are those which become labeled and these only represent a small portion of the polysaccharide chains. Therefore, little is known about the actual acceptor molecules or the endogenous polysaccharides. Villemez (19) also found that GDP-man stimulated the incorporation of GDP-^{14}C-glu into alkaline-insoluble material.

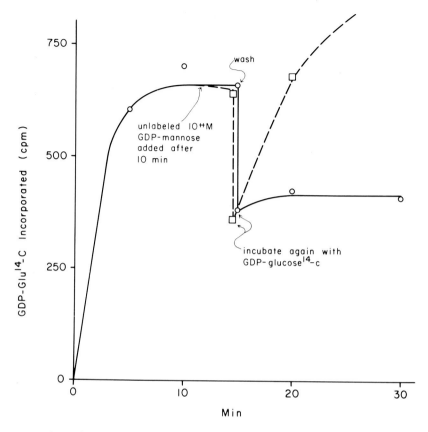

Fig. 9 Effect of preincubating enzyme fraction in GDP-man on the incorporation of GDP-glu-^{14}C into an alkali-insoluble polysaccharide fraction.

However, these workers found that GDP-man did not stimulate the initial rate of incorporation but in-creased the total amount of incorporation by prolong-ing the time of incorporation. That is the reaction proceeded for a longer period in the presence of GDP-man. These results are in agreement with the idea that mannose increases the number of acceptor sites, or provides new acceptor sites, for glucose attachment. However, whether these sites are part of the cellu-lose molecule or are part of some other polysac-charide is not known. Heller and Villemez (20) have recently solublized the enzyme which catalyzes the incorporation of mannose into polysaccharide. The enzyme which incorporates GDP-glu into polysaccharide has also been solubilized by Liu and Hassid (21). With the use of these solubilized transferases it should be possible to study the nature of the acceptor molecules and to determine the relationship of these various reactions to cellulose and cell wall biosynthesis.

Cellular Location of the Various Glucosyl Transferases

In order to determine the cellular location of the UDP-glu and GDP-glu:polysaccharide transferases and to determine whether these activities resided in the same fraction, the particulate enzyme fraction from cotton fibers was subjected to isopycnic density gradient centrifugation. Fractions were collected from the centrifuge tube and were assayed for the incorporation of UDP-^{14}C-glu and GDP-^{14}C-glu into polysaccharides as well as the transfer of glucose from UDP-glu to β-sitosterol. As shown in Figure 10, all of these enzymatic activities were found to be located in the same fraction which had an equilibrium density comparable to 25% sucrose. Thus, the β-(1→ 3) and β-(1→ 4) glucan synthesizing enzymes and the steroid glucoside activity all appear to be located in a similar membranous fraction. In addition to the activities shown in Figure 10, the same sucrose denstiy fraction has been shown to catalyze the transfer of mannose from GDP-man-^{14}C into both a glycolipid and a polysaccharide. The transfer of glucose from GDP-glu by this enzyme fraction was also stimulated by GDP-man. Not only did this enzyme fraction contain the above enzymes but it apparently contained the acceptor molecules as well. To determine the nature of the enzyme fraction, it was examined by electron microscopy. As shown in Figure 11, a large number of vesicular-like bodies were observed indicating that the enzymes

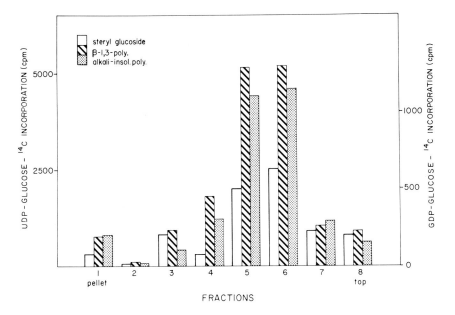

Fig. 10 Isopycnic sucrose density (20-45%) gradient centrifugation of particulate enzyme fraction. ▦ GDP-glu-^{14}C incorporation into alkaline-insoluble polysaccharide; ◫ UDP-glu-^{14}C incorporation into alkaline soluble polysaccharide; ☐ UDP-glu-^{14}C incorporation into steryl glucoside.

are located on membranous structures. However, whether these vesicles arise from plasma membranes or whether they represent fragments of Golgi or endoplasmic reticulum is not know. Ray et al reported a UDP-glu:polysaccharide glucosyl transferase from peas which catalyzed the synthesis of a cellulose-like polysaccharide (22). These workers separated the enzyme by sucrose density gradient centrifugation and identified the active enzyme fraction as being part of the Golgi apparatus.

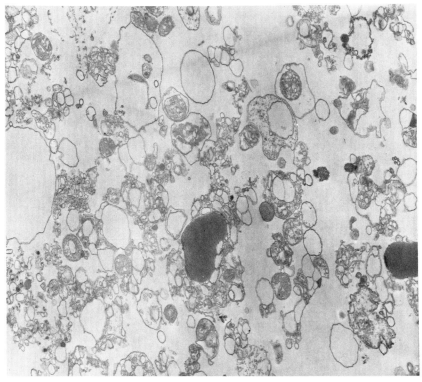

Fig. 11 Electron micrograph of the membrane vesicles isolated by sucrose density gradient centrifugation which actively synthesize polysaccharides and glycolipids.

Possible Intermediates in the Biosynthesis
of Cell-Wall Polysaccharides

During studies on polysaccharide synthesis in cotton fibers, it was found that considerable radioactivity was transferred from GDP-^{14}C-man into chloroform-methanol soluble products. The ^{14}C-lipid formed by the particulate enzyme was quite unusual in that it was unstable in these incubations and appeared to turn over. Thus the lipid had properties of an intermediate compound rather than

of a structural material. Certain types of glycolipids have been found to be intermediates in the
biosynthesis of bacterial cell-envelope polysaccharides. In order to determine whether this mannolipid might be similar to those described in bacteria,
the synthesis and stability of the lipid was followed with the time as shown in Figure 12. The incorporation of mannose into the lipid was linear for
about 10 minutes and then leveled off. At the end
of 15 minutes, the particulate enzyme was isolated
by centrifugation and washed to remove GDP-man and
then reincubated in the absence or presence of
additional GDP-^{14}C-man. When GDP-man was added in
the second incubation, radioactivity was again incorporated into the lipid. However, in the absence
of GDP-man, radioactivity in the lipid began to
decline and fell about 30% in 15 minutes. Incorporation into polysaccharide was also followed, but
the radioactivity disappearing from the lipid did
not appear to be incorporated into the polysaccharide.
However, this may be due to the fact that radioactivity was too low to detect. Since the lipid
did appear to be turning over, it was decided to
study the nature of this material to see whether it
had properties similar to known intermediate glycolipids.

The mannolipid synthesized by cotton fibers was
isolated by extraction with chloroform-methanol and
was purified by chromatography on DEAE-cellulose.
Over 90% of the radioactivity was retained on DEAE-
cellulose and was eluted with ammonium acetate.
The radioactive peak was treated with NaOH to saponify phospholipids and then was further purified on
DEAE-cellulose. Again, the bulk of the radioactivity
(95%) was retained on this column and was eluted in
a symmetrical peak with a linear gradient of
0 to 0.05 M ammonium acetate (Figure 13). Radioactive fractions were pooled and concentrated and
this purified mannolipid was used for chemical
characterization. As shown in Figure 14, the manno-

283

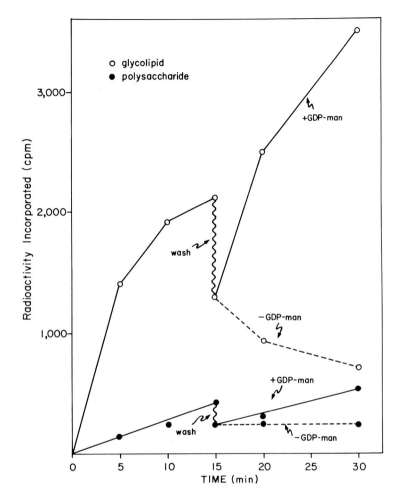

Fig. 12 Incorporation of GDP-^{14}C-man into glyco-lipids and alkali-insoluble polysaccharides by a par-ticluate enzyme from cotton fibers. After incubation for 15 min with GDP-^{14}C-man, the particulate enzyme was isolated by centrifugation, washed and resuspend-ed in buffer. One half of the resuspended enzyme was incubated in the absence (----) of GDP-^{14}C-man and the other half in the presence of GDP-^{14}C-man (———). Lipids were extracted with chloroform-methanol and insoluble polysaccharide was isolated by centrifuga-tion and extracted with alkali.

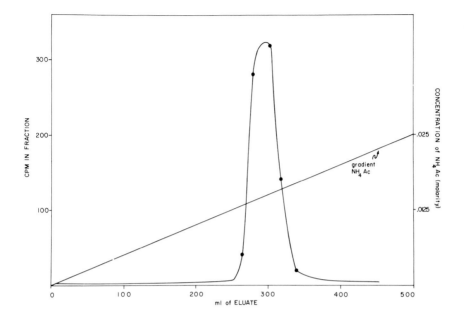

Fig. 13   Purification of ^{14}C-mannolipid on DEAE-cellulose.   Lipids were eluted with a linear gradient of 0 - 0.05 M ammonium acetate.

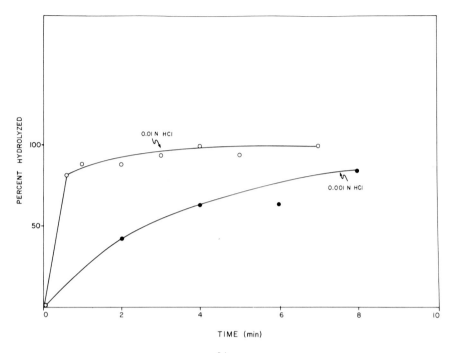

Fig. 14   Release of ^{14}C-mannose from mannolipid
by mild acid hydrolysis. Mannolipid was placed in
0.01 N HCl or 0.001 N HCl in 50% propanol and heated
at 100° for the times indicated. Mannose and lipid
were partitioned in chloroform:methanol:water.

lipid was very labile to mild acid hydrolysis. Thus,
all of the radioactivity was released into the
aqueous phase in 1 min or less when the lipid was
placed in 0.01 N HCl at 100°. In 0.001 N HCl about
50% of the radioactivity became water-soluble in
about 3 or 4 min. The only radioactive material released by this treatment was identified as mannose
by paper chromatography in several solvent systems.
In addition, the mannolipid had a hexose to phosphate
ratio of about 1.

The mannolipid was subjected to thin layer
chromatography in three different solvent systems
(neutral, acidic and basic) as shown in Table V.

286

TABLE V

Thin Layer Chromatographic Mobility of Glycolipids
Synthesized from GDP-Mannose-^{14}C and UDP-Glucose-[14]C

	Rf's in		
	Solvent I (acidic)	Solvent II (neutral)	Solvent III (basic)
Man-Lipid	0.88	0.35	0.18
Glu-Lipid	0.89	0.42	0.21

Solvent I   Chloroform:Methanol:Acetic Acid:H_2O (25:15:4:2)
Solvent II   Chloroform:Methanol:H_2O (65:25:4)
Solvent III Chloroform:Methanol:HH_4OA (75:25:4)

The lipid exhibited mobilities (23) similar to those
of a mannosyl-phosphoryl-polyprenol which is an
intermediate in bacterial cell-wall polysaccharide
synthesis (24). Probably the most significant finding
which suggests that this lipid may represent an
activated form of mannose and therefore could serve
as an intermediate is the finding that synthesis
of the mannolipid is reversed by GDP but not by GMP.
As shown in Table VI, the particulate enzyme was
incubated with GDP-^{14}C-man to allow synthesis of the
mannolipid. The particulate enzyme was isolated by
centrifugation, washed to remove GDP-man and then
resuspended either without additions or in the presence of GDP or GMP. It can be seen that in the control or in the presence of GMP all of the radioactivity
remained in the mannolipid but when GDP was added
radioactivity was transferred from the lipid to GDP to
form GDP-^{14}C-man. This experiment demonstrates that
the synthesis of the mannolipid is a reversible
reaction and that the lipid can actively donate its
mannose moiety. Thus far, we have not been able to
identify the lipid moiety of the mannolipid. We have
attempted to characterize the lipid portion of the

TABLE VI

Transfer of ^{14}C-Mannose From Mannolipid
to GDP

Additions to reaction mixture		CPM in CHCl$_3$:MeOH			CPM in GDPM		
		0 min	1 min	5 min	0 min	1 min	5 min
Exp. 1							
	None	2030	2100	2010	146	180	174
	GMP (5 X 10^{-5}M)	–	2190	1975	–	166	–
	GDP "	–	1100	800	–	–	617
	GTP "	–	1175	900	–	560	–
Exp. 2							
	None	4304	–	3934	173	–	167
	GMP (5 X 10^{-4}M)	–	3832	3911	–	168	186
	GDP	–	2671	2411	–	808	963

In this experiment, the particulate enzyme was incubated
with GDP-^{14}C-man for 10 min and was then centrifuged to isolate
the enzyme and washed to remove GDP-man. The enzyme was
resuspended, divided into equal aliquots and incubated either
alone, or with GMP, GDP or GTP. Zero time refers to the start
of the second incubation. Disappearance of radioactivity from
chloroform-methanol and appearance of radioactivity in GDP-man
(isolated by paper chromatography) was then followed. It should
be pointed out that most of the mannolipid remains associated
with the enzyme fraction unless extracted with chloroform-methanol.

mannolipid by mass spectrometry but have not been able
to draw any conclusions about the nature of this
material. That is, the lipid does not show a spectrum
typical of polyprenols as do the bacterial intermediate

lipids, but whether this is due to contamination by other lipids or to the fact that the lipid has a different structure cannot be determined at this time.

In order to gain additional information about the nature of the lipid itself it was desirable to obtain an enzyme fraction free of lipid acceptor so that an active lipid acceptor could be isolated. Therefore the particulate enzyme was treated with acetone (25) to remove endogenous lipids and this enzyme preparation was used for the synthesis of the mannolipid. As shown in Figure 15, the acetone-treated enzyme demonstrated no activity in the absence of exogenous lipid acceptor. An active lipid acceptor was isolated by extraction of cotton bolls with chloroform-methanol and purification on DEAE-cellulose. The active lipid was eluted from DEAE-cellulose with ammonium acetate and emerged just after the mannolipid. This material contained phosphorus and is currently being examined to determine the chemical composition of the lipid. Attempts to determine the ultimate acceptor of mannose from the mannolipid have not been successful but it does seem likely that this mannolipid is an intermediate in the synthesis of some cell-wall polymer.

We have also found that the same particulate enzyme incorporated glucose from UDP-^{14}C-glu into chloroform-methanol soluble products. The bulk of the radioactivity in the chloroform-methanol fraction is incorporated into steroid glucosides but a significant amount of radioactivity is found in an acidic lipid which has properties similar to the mannolipid. Thus, after isolation on thin-layer plates and purification by DEAE-cellulose chromatography the ^{14}C-glucolipid had R_f's similar to those of the mannolipid in three different solvent systems (Table V). The glucolipid was very labile to acid

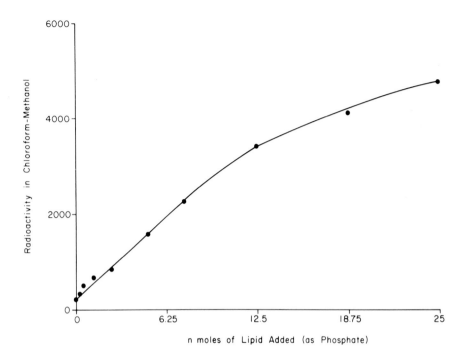

Fig. 15 Effect of exogenous lipid acceptor on the incorporation of GDP-^{14}C-man into chloroform: methanol. Enzyme was treated with acetone to remove endogenous lipids and then was assayed with lipid acceptors.

---

hydrolysis just as was the mannolipid. Thus, 50% of the glucose was released in about 2 min at 100° in 0.003 N HCl. Finally, as shown in Table VII, the synthesis of the glucolipid was reversible. The particulate enzyme was incubated with UDP-^{14}C-glu for 8 minutes and was then centrifuged and washed to remove any remaining UDP-glu. The particulate enzyme was then resuspended and incubated either alone or in the presence of UMP or UDP. It can be seen that in the presence of UDP, radioactivity disappeared from the chloroform-methanol phase and

TABLE VII

Transfer of ^{14}C-Glucose from
Glucolipid to UDP

Additions to 2nd Incubations	Incubations Time (min)	CPM in Glucolipid	CPM in UDP-Glu
NONE	0	630	56
NONE	5	559	60
UMP (5 X 10^{-3}M)	5	581	62
UDP "	5	239	410

Conditions were as outlined in Table VI for the mannolipid
except in this case UMP or UDP were used to test for reversal.

was found in UDP-glu. However, in the absence of
UDP (i.e., no additions or UMP addition), there was
no transfer of radioactivity from the glucolipid to
the aqueous phase (hence, no formation of UDP-glu).

In order to demonstrate that these glycolipids
were present in the intact cotton bolls and were not
the result of artifacts, a large number of cotton
bolls were extracted with chloroform-methanol and
the lipid fraction was purified by DEAE-cellulose
chromatography. The acidic lipids which eluted
with ammonium acetate were treated with NaOH to
saponify phospholipids and then were further purified by DEAE cellulose chromatography. Before the
last purification some of the ^{14}C-mannolipid was
added as a marker and the radioactive peak was
isolated. This lipid fraction was subjected to
mild acid hydrolysis (0.01 N HCl, 10 min, 100°) to
isolate the sugars and these sugars were reduced
with NaBH$_4$ and converted to their acetate derivatives
for gas liquid chromatography. Figure 16 shows a

291

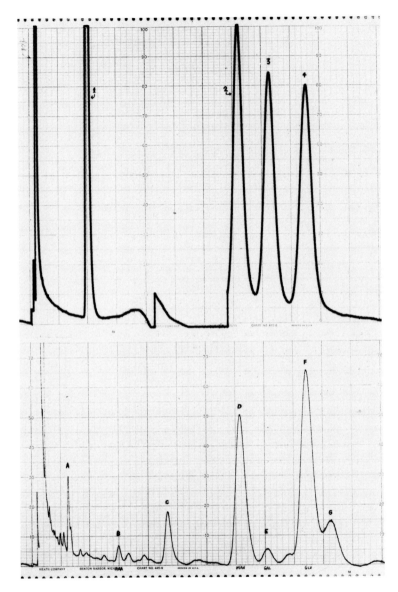

Fig. 16   Gas liquid chromatography of the sugars released from the acidic lipid fraction by mild hydrolysis.   Upper tracing is of standards as follows: 1 = sugar alcohol acetate of fucose, 2 = mannose, 3 = galactose, 4 =glucose.   Lower tracing = alcohol acetates for the acidic lipid fraction.

tracing of the gas-liquid chromatogram of the acetylated sugar alcohols obtained from the endogenous
lipids. Two major peaks were observed (D and F)
which correspond to mannose and glucose based on the
migration of known standards. Smaller peaks corresponding to galactose (E) and arabinose (B) were also
detected. However, peaks A, C and G have not yet
been identified. This experiment indicates that a
number of sugars besides glucose and mannose may
exist attached to lipids by an acid-labile linkage.

Although the role of those compounds has not
yet been determined, the similarity of these lipids
to those isolated from bacterial systems and shown
to be intermediates in cell wall synthesis is
striking. In bacterial systems, the lipids are believed to act as carriers of hydrophilic sugar molecules through the hydrophobic membrane to the region
of the cell walls. A similar analogy could be used
in the case of the plant cell walls. In fact, the
first suggestion of glycolipid type intermediates
in polysaccharide synthesis was made by Colvin more
than 10 years ago (26). He presented preliminary
evidence for the existence of a glucolipid as an
intermediate in cellulose synthesis by the bacterium
Acetobacter xylinum. Since that time other workers
have found preliminary evidence for the existence of
glycolipids in higher plants. Thus Villemez and
Clark (27) found evidence for particle bound lipid-
like intermediates of glucose and mannose in mung
bean seedlings but were not able to obtain sufficient
amounts for characterization. Alam and Hemming (28)
synthesized betulaprenol phosphate and showed that
this compound was an acceptor of mannose from GDP-[14]C-
man with a particulate enzyme preparation from mung
beans. Thus, the plant systems may contain similar
types of precursors but more data is needed before
this hypothesis can be verified.

This work was supported by the Robert A. Welch
Foundation and Cotton Incorporated.

References

1. M. J. Rollins, *Forest Prod.* *18*, 91 (1968).
2. P. Albersheim in *Plant Biochemistry* (Eds. J. Bonner and J. E. Varner) Academic Press, New York, p. 151 (1965).
3. F. Shafizadeh and G. D. McGinnis *Advan. Carbohyd. Chem. Biochem.* *26*, 297 (1971).
4. H. E. Dadswell, H. G. Wardrop, *World Forestry Congr. 5th, Seattle, Wash.* *2*, 1279 (1960).
5. G. A. Adams, and C. T. Bishop *Tappi* *38*, 772 (1955).
6. B. G. Ramby, and B. Grinsberg, *Compt. Rend.* *236*, 1402 (1950).
7. A. D. Elbein, G. J. Barber, and W. Z. Hassid, *J. Amer. Chem. Soc.* *86*, 309 (1964).
8. S. Roseman, J. Distler, J. Moffatt, and H. Knorana, *J. Amer. Chem. Soc.* *83*, 659 (1961).
9. G. A. Barber, A. D. Elbein, and W. Z. Hassid, *J. Biol. Chem.* *239*, 4056 (1964).
10. D. O. Brummond, and A. P. Gibbons, *Biochem.* *324*, 308 (1965).
11. L. Ordin and M. A. Hall, *Plant Physiol.* *42*, 205 (1967).
12. K. K. Batra and W. Z. Hassid, *Plant Physiol.* *44*, 755 (1969).
13. H. M. Flowers, K. K. Batra, J. Kemp and W. Z. Hassid *Plant Physiol.* *43*, 1703 (1968).
14. C. Lenoel and M. Axelos *FEBS Letters*, *8*, 224 (1970).
15. D. S. Feingold, E. I. Neufeld and W. Z. Hassid, *J. Biol. Chem.* *233*, 783 (1958).
16. J. Chambers and A. D. Elbein, *Arch. Biochem. Biophys.* *138*, 670 (1970).
17. G. Franz and H. Meier, *Phytochemistry* *8*, 579 (1969).
18. A. D. Elbein, *J. Biol. Chem.* *244*, 1608 (1969).
19. C. L. Villemez, *Biochem. J.* *121*, 151 (1969).
20. J. S. Heller and C. L. Villemez, *Biochem. J.* *128*, 243 (1972).

21. Tin-Yin Lui and W. Z. Hassid, *J. Biol. Chem.* *245*, 1922 (1969).
22. P. M. Ray, T. L. Shiniger and M. M. Ray, *Proc. Nat. Acad. Sci. U.S. 64*, 605 (1969).
23. J. W. Baynes and E. C. Heath, *Fed. Proc. 31*, 437 (1972).
24. M. Scher, W. F. Lennarz, and C. C. Sweeley, *Proc. Nat. Acad. Sci. U.S. 59*, 1313 (1968).
25. A. T. Frederic, F. E. Frerman and E. C. Heath *J. Biol. Chem. 246*, 118 (1971).
26. J. R. Colvin, *Nature 183*, 1135 (1959).
27. C. L. Villemez, and A. F. Clarkk, *Biochem. Biophys. Res. Commun. 36*, 57 (1969).

AN INDOLE-3-ACETIC ACID ESTER OF
A CELLULOSIC GLUCAN

Robert S. Bandurski* and Z. Piskornik**
Department of Botany and Plant Pathology
Michigan State University
East Lansing, Michigan   48823

Abstract

A bound auxin of *Zea mays* has been purified and partially characterized. It is a high molecular weight, lipophilic cellulosic glucan containing indole-3-acetic acid in ester linkage. This compound accounts for one-half of the total indole-3-acetic acid of mature sweet corn kernels.

Symbols

IAA = indole-3-acetic acid
TMS = trimethylsilyl
GLC = gas-liquid chromatography
TLC = thin layer chromatography

Introduction

The inclusion of this topic in a symposium devoted to the biogenesis of plant cell wall

---

*Report of work supported, in part, by the U.S. National Science Foundation GB-18353-X

**Polish American Agricultural Exchange Program – Present Address, Plant Physiology Department, Agricultural University, Cracow, Poland.

polysaccharides is reminiscent of Bob Newhart's saga
of the submarine, the U.S.S. Codfish. Gentlemen,
in a moment we will be surfacing and you will gaze
on the skyline of either New York-or-or-Buenos Aires?
Similarly as we look back on this symposium 10 years
hence, we will either be amazed at the sagacity of
Dr. Loewus in keeping us right on track or dismayed
that he led us far astray. I will try though to
provide a bit of rational for the inclusion of this
data on an IAA-glucan -- including my thoughts on
why conjugation of IAA with various carbohydrates
may have importance in cell wall metabolism.

General Considerations

The primary cell walls of plants undergo enlargement during growth and growth is under hormonal
control. Thus, cell wall enlargement is hormonally
controlled. Two types of hormonal control of wall
enlargement may be envisaged -- direct or indirect.
In a direct control mechanism, one might postulate
that the hormone affects the wall, possibly by
permitting leakage of wall-softening enzymes into the
wall, or by acting as a co-factor or activator of
wall-softening enzymes. The remainder of the growth
syndrome would then be a consequence of wall-softening. That is, the cell would distend with water
stretching the softened wall. This would disturb
the cytoplasmic homeostatic mechanisms, and enhanced
respiration, protein, RNA and DNA synthesis would
ensue. Fig. 1, locus B, depicts a system where the
wall is the primary target of the hormone and the
metabolic machinery is only subsequently activated.
You will note that the metabolic machinery is
schematisized as a series of interlocking and rigidly
coupled gears. This rigid coupling is due to the
fact that each major pathway is feedback controlled
(the arrows on the gears) and each pathway depends
upon another pathway for substrate. For example, if
the carbohydrate system stops, the organic acid system will also stop due to a lack of precursors and

298

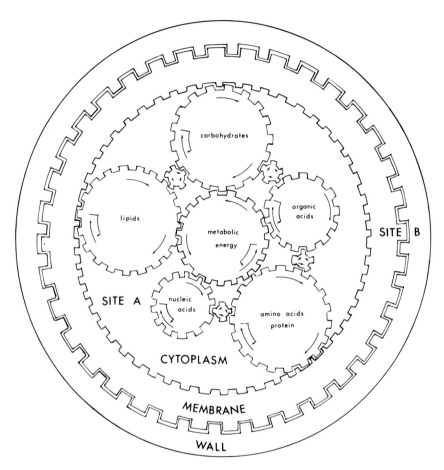

Fig. 1 A diagrammatic representation of two possible sites of auxin action leading to cell wall changes. Site B represents a direct action of auxin on the wall (as an enzyme cofactor) or on the membrane (as a permeability affector). Site A represents an indirect action of auxin on cytoplasmic metabolism which indirectly results in cell wall changes.

this, in turn, will stop amino acid biosynthesis and protein synthesis. Thus growth is essentially an all or none process.

An indirect control system could also lead to changes in the primary wall. For example, the hormone might act as an allosteric regulator of energy utilizing systems. In the Atkinson sense (1) the hormone could slightly lower the ATP/ADP ratio at which synthetic enzymes would function. This switching on of the biosynthetic enzymes might lead to changes in membrane permeability and finally to leakage of wall-softening enzymes into the wall. Activating the metabolic machinery of the cytoplasm would then be the primary target of the hormone and wall-softening would be a consequence. This system is represented diagramatically in Fig. 1 as locus A.

Present data are insufficient to decide between these alternatives. The point that is relevant to this symposium is that *directly or indirectly hormonal regulation of growth results in changes in the physical, and possibly chemical, properties of the primary cell wall*. Thus, it may be of significance, that in the kernels of *Zea mays* the plant growth hormone, IAA, occurs primarily, if not entirely, in ester linkage with various carbohydrates. In particular we will be concerned with the isolation and properties of a compound containing IAA and a cellulosic glucan.

Results

The flow sheet of Fig. 2 illustrates the manner of isolation of the indolylic compounds of the kernels of *Zea mays* (2). The ground corn meal is extracted with aqueous acetone. This solubilizes the small amount, if any, of free IAA and *all* of the alkali-labile esters (3). Non alkali-labile indole compounds that do not react with Salkowski reagent and are not soluble in aqueous acetone would be undetected by our methods.

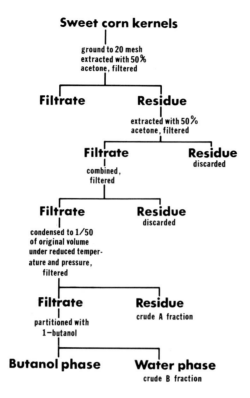

Sweet corn kernels
|
ground to 20 mesh
extracted with 50%
acetone, filtered

Filtrate   **Residue**
|
extracted with 50%
acetone, filtered

Filtrate   **Residue**
| discarded
combined,
filtered

Filtrate   **Residue**
| discarded
condensed to 1/50
of original volume
under reduced temper-
ature and pressure,
filtered

Filtrate   **Residue**
| crude A fraction
partitioned with
1-butanol

Butanol phase   **Water phase**
                    crude B fraction

Fig. 2 Flow sheet illustrating the isolation
procedure for the IAA-glucan (A fraction) and the
IAA-cyclitol (B fraction).
(From A. Ehmann and R. S. Bandurski, *J. Chromatog*.,
72, 61 (1972).

The A fraction is that fraction which precipi-
tates from the acetone-water extract as the acetone
is evaporated *in vacuo* and the system becomes aqueous.
It is this fraction which contains IAA-glucan. The
B fraction is that which remains in the aqueous phase.
It contains the compounds shown in Fig. 3.

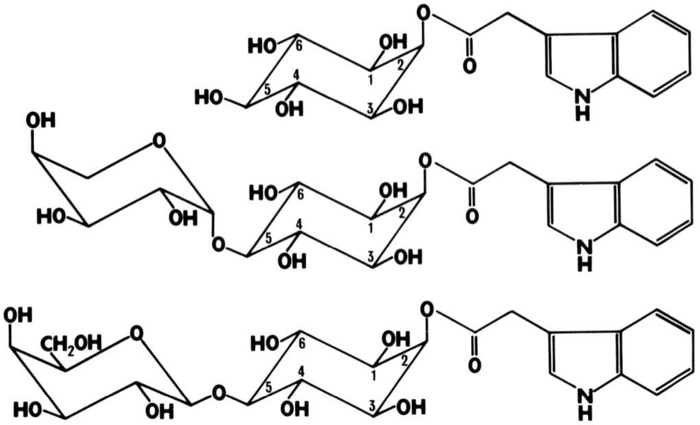

Fig. 3 Representative compounds of the B fraction (Courtesy of A. Ehmann).

A few words about these compounds. Dr. Hamilton in our laboratory in 1961 (4) noted that free IAA could be isolated from growing young corn seedling tissue only after alkaline hydrolysis of the tissue extracts or after ether-induced tissue autolysis. This initiated our interest in IAA esters and turned us toward a study of the IAA esters of corn kernels where they were present in greatest amount. In the kernel tissue we found small amounts of free IAA but confirmed the original observations of Berger and Avery (5) and Haagen-Smit et al., (6) that the bulk of the IAA was released by alkaline hydrolysis and thus might be in ester linkage. None of these "bound-IAA" compounds were chemically characterized. During the period 1964 to 1968, first Labarca, et al., (7) then Nicholls (8) and later Takano et al., (9), Udea et al., (10), and most recently, Ehmann (11, 12) succeeded in isolating and characterizing the esters. It is worthy of note that they all proved to be esters and that the alcohol was either *myo*-inositol, a *myo*-inositol glycloside or a hexose (A. Ehmann, unpublished). To the best of our knowledge this is the

302

first time that all, or perhaps, almost all, of the
indolylic compounds of a plant tissue have been
identified.

The data of Table I show some of Dr. Ueda's

Table I. Quantitative Estimation of IAA Compounds in Corn Kernels

The value for free IAA has been corrected for losses during analysis using

the experimentally determined factor of 1.27. The values for all other

compounds are not corrected.

	1963 Harvest		1967 Harvest	
	mg IAA/kg	Percent	mg IAA/kg	Percent
Total	88	(100)	73	(100)
Free IAA	10	12	0.5	0.7
A group	35	38 (100)	42	58 (100)
B group	43	49 (100)	30	41 (100)

After: Ueda, M., Bandurski, R.S. Plant Physiol. 44, 1175 (1969).

work on the relative amounts of the A (high molecular
weight, water insoluble) and B (low molecular weight,
mainly water soluble) compounds. Approximately
one-half of the total alkali-labile IAA compounds are
in the A group and one-half in the B group. Recent
work by Dr. Piskornik (unpublished) indicates that
almost all of the alkali-labile IAA compounds are
in the endosperm of the corn kernel (about 10^{-3} M)
with much lesser amounts in the embryo *plus scutellum*.
For example, kernels germinated for 48 hours were
separated into endosperm and embryo plus scutellum
tissue. A total of 437 grams, dry weight, of endosperm tissue was found to contain 12.4 mgs of alkali
releasable IAA in the B fraction and 17.7 mgs of
alkali-releasable IAA in the A fraction. This amounts

to 40.5 and 28.4 mg/kg of A and B fraction alkali-
-releasable IAA respectively and a total of 68.0 mg
of alkali-releasable IAA per kilogram of dry endosperm. In this same experiment 78.2 grams of scutellum plus embryo tissue was collected. This tissue
was ground and extracted with acetone-water in the
same manner as that used for endosperm tissue. The
concentration of indolylic compounds was, however,
too low for satisfactory quantitative estimation.
From prior work (10), we would estimate the IAA content at about one 1/100th of the concentration that
occurs in endosperm tissue. Characterization of the
small amount of indolylic compounds present in growing vegatative tissue remains a task for the future.

Unlike the B compounds, which are of low molecular weight and therefore amenable to structural
studies using combined GLC-mass spectrographic
methods, the "A" or glucan fraction has proven re-
fractory. At present we know only that the fraction
contains IAA and a cellulosic glucan. The purification of this fraction is as follows:
    Pre-purification. The zein-like sticky yellow
A fraction is dissolved in aqueous ethanol, adjusted
to pH 8, and impurities removed by ether extraction.
The aqueous phase is then adjusted to pH 2.5 and the
ether extraction repeated. The A fraction precipi-
tates from solution and is dried by lyophylization.
Alternatively, a higher molecular weight A fraction
can be obtained by repeatedly dissolving crude A
fraction in a minimal volume of ethanol-water and the
A fraction precipitated by chilling the solution to
-20 C collected and dried by lyophylization.
    Silica gel column chromatography. Fractionation
of the A compound into a fraction elutable with
chloroform-methanol and by an ethanol-water gradient
has been previously described (2). Fig. 4 shows the
elution profile obtained when crude (not pre-purified)
A fraction is absorbed onto silica gel and then
sequentially eluted with chloroform, chloroform-ace-
tone, acetone and finally an acetone-water gradient.

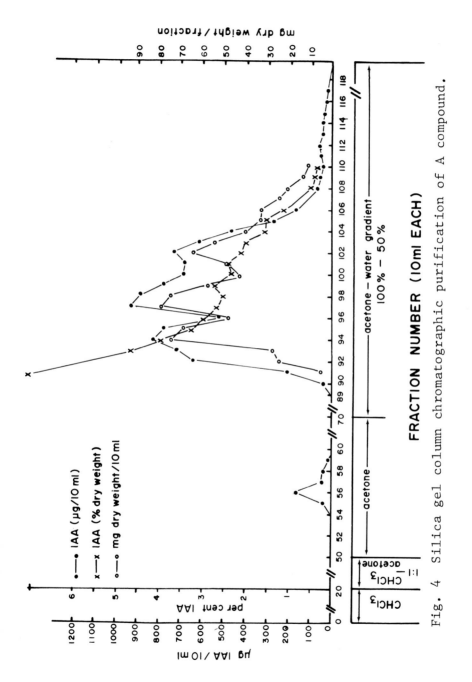

Fig. 4 Silica gel column chromatographic purification of A compound.

Traces of free IAA, present in crude A, are removed
in the acetone wash. The column gives good separation of free IAA and bound alkali-labile IAA. The A
fraction elutes in several poorly resolved peaks.
We have not detected qualitative differences in the
hydrolysis products obtained and believe the multiple
peaks are due to variations in length of the glucan
chain.

Lipophylic Sephadex Chromatography. Data indicating that the A compound is a polymer of variable
chain length is shown in Fig. 5. For this experiment the A fraction eluted from a silica gel column
was separated into two fractions, the leading peak
called Z-1 and the trailing peak, Z-2. Z-1 is that
eluted with acetone containing a small amount of
water and Z-2 is that fraction eluted with acetone
containing more water. Z-1 and Z-2 were each then
chromatographed on lipophylic Sephadex, (Fig. 5 A
and B). As can be seen Z-1 has a peak elution volume
of 18 ml and Z-2 of 10 ml. Z-1 thus is of lower
molecular weight than Z-2 although qualitatively
similar in hydrolysis products (see below). The
thin layer chromatograms of the Sephadex eluent
confirm the conclusion from the Sephadex column.
Using a non-polar solvent, Z-1 fractions have a higher
TLC migration than the Z-2 fractions and for both
fractions there is a gradation in mobilities with
the lowest R_f for low retention time Sephadex eluents
and a high R_f for long retention time Sephadex
eluents. This is the behavior to be expected for a
polymeric compound of variable chain length.

Thin layer chromatography. The streaking observed on this layer chromatograms is not attributable
to decomposition of the compound during chromatography. Fig. 6 shows a TLC plate on which un-frac-
tionated A compound is chromatographed. A streak
results from the origin to an R_f of about 0.6. If
now sectors of the streak are eluted and re-chromato-
graphed, each sector moves again to the same R_f as

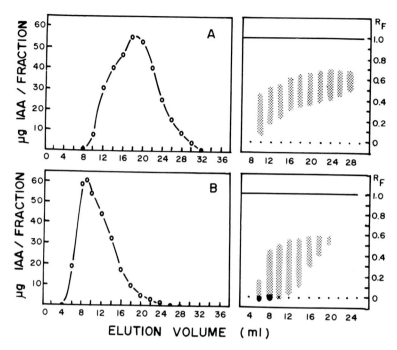

Fig. 5 Elution profiles and TLC patterns of the A compounds from an LH-20 Sephadex column. A: Z-1 type of A compound (lower molecular weight); B Z-2 type of A compound (higher molecular weight). Each fraction was collected separately and chromatographed on pre-coated silica gel plates using ethyl acetate-methyl ethyl ketone-ethyl alcohol-water solvent (5:3:1:1). Detection was with Ehrlich reagent. (From Z. Piskornik and R. S. Bandurski, *Plant Physiol*. *50*, 176 (1972).

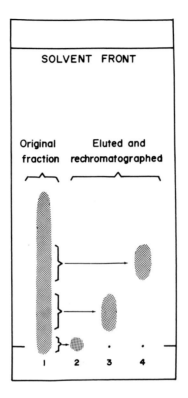

Fig. 6 Tracing of a silica gel chromatogram showing mobility of (1) purified A fraction and (2), (3) and (4) the mobility of A after elution from the indicated sector and rechromatographing. The solvent was as for Fig. 5 and visualization was by charring.
(From Z. Piskornik and R. S. Bandurski, *Plant Physiol.* *50*, 176 (1972).

that from which it had been eluted.

The coincidence of Ehrlich reactive indolylic material with sulfuric acid charable substance is a major reason for believing the A compound to be of high purity. Dr. Piskornik has tested some 50 chromatographic solvents which gave very different TLC mobilities to the A compound. In every case there has been a complete coincidence of Ehrlich reactive indolylic compounds with organic matter visualizable by charring. Further, partial hydrolysis of A Compound by crude or homogeneous cellulase (2) markedly increases the TLC mobility of the A compound. Again, when this partially degraded compound is chromatographed there is complete coincidence between Ehrlich reactivity and charrable matter.

Solubility. The A compound is insoluble in water and in all of the common organic solvents, such as chloroform, hexane, benzene, ethyl acetate, diethyl ether, absolute alcohol and dry acetone. The compound is soluble in aqueous alcohol or acetone particularly in the range of 70 to 90% alcohol or acetone.

IAA content of the A compound. Following alkaline hydrolysis (15 min, 22 C, 1.0 N NaOH) IAA is liberated and may be estimated quantitatively by its UV-absorption or its reactivity in the Salkowski test (2). The IAA has been identified by its biological activity and paper chromatographic properties (4), by its TLC mobility (2) and by gas liquid mass spectrometric analysis of the trimethysilyl derivatives. Derivatization was with Bis(trimethylsilyl)trifluoroacetamide containing 1% trimethylchlorosilane in N, N- dimethylformamide for 1 hr at 55 C. Under the silylation conditions used 2 TMS groups are introduced yielding a compound with a molecular ion at m/e = 319. If ammonium hydroxide is used for alkaline hydrolysis approximately equal amounts of IAA (m/e = 319) and IAA amide (m/e = 318) are formed.

The mass fragmentation patterns of authentic IAA
and IAA amide are identical to those obtained from
putative IAA and IAA amide obtained from the A compound. Formation of indoleacetamide upon ammonoly-
sis of A compound shows that the IAA is in ester
linkage to the compound.

The IAA content, as percent of the dry weight,
of A compound is shown graphically in Fig. 4. As
can be seen the percent IAA ranges from 7% to less
than 1%. Assuming one mole of IAA per mole of A
compound, one may estimate a molecular weight of
2500 to 17,500. Since (see below) the glucose content upon hydrolysis is about 40 to 50%, the number
of glucose residues per molecule would be of the
order of 7 to 50.

Strong acid hydrolysis of alkali-hydrolyzed A
compound. Following alkaline hydrolysis and removal
of ether-soluble constituents, the residual water
soluble material was hydrolyzed for 3 hours with 1
N trifluoroacetic acid. After hydrolysis, ionic
material was removed with a mixed bed cation-anion
exchange resin and the resultant solution analyzed
for glucose. No sugars other than glucose were detected by GLC or TLC and there was excellent agreement between the estimates of glucose as reducing
sugar and as glucose by the highly specific glucose
oxidase assay. In general the Z-1 type compounds
contained, respectively, 34 to 36% and 31 to 35% of
glucose as estimated by the Nelson reducing sugar
assay and by the glucose oxidase method. Z-2 type
compound contained 46 to 48 and 44 to 48% glucose as
estimated by the two methods. Thus apparently one-
third to one-half of the weight of the hydrolyzed
compound is accounted for as glucose.

Partial acid hydrolysis of alkali-hydrolyzed
A *compound.* By shortening the time of acid hydrolysis to 30 min at 100 C in 1 M trifluoroacetic acid,

it was possible to obtain a series of cellodextrins as hydrolysis products of the A compound. The hydrolysates were separated by Sephadex G-15 filtration into a monosaccharide, disaccharide, trisaccharide, tetrasaccharide and higher oligosaccharide fraction. Fig. 7 shows the TLC mobilities of hydrolysis products from the A compound glucan together with appropriate standards. TLC and GLC mass spectrographic examination disclosed only glucose, cellobiose, cellotriose, cellotetrose, and higher cellodextrins as hydrolysis products. No other di- or trisaccharides, including mixed 1→3, 1→4 trisaccharides were detected. The identity of the cellobiose and cellotriose was confirmed by comparison of the mass spectrographic fragmentation patterns of the TMS ethers of the putative cellobiose and cellotriose with those of authentic cellobiose and cellotriose.

Unidentified hydrolysis products. A purified Z-2 type A compound containing 2% IAA was found to contain 2.44% N. This corresponds to about 15 moles of N per mole of IAA after correcting for the N content of the IAA. Impure preparations of A compound contain traces of two peptides resolvable by TLC. Upon acid hydrolysis of the peptide aspartate, serine, glutamate, proline, glycine, alanine and leucine are detectable. None of the amino acids are present in amounts exceeding 0.05 moles/mole of IAA. We thus have not accounted for the N content of the A compound.

The unidentified substituents are, at least, three in number. They are not hydrolyzed from the A compound by alkali but are liberated by acid hydrolysis. The compounds are detectable by iodine vapor, by fluorescence in long wave UV after spraying with alkali, by permanganate or by the Pauly reagent. Our working hypothesis is that they are unstable polyhydric amines linked to the glucan by a glycosidic bond.

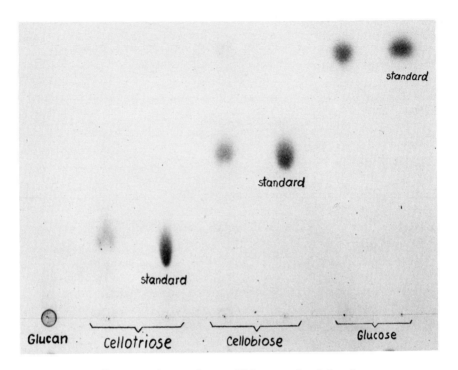

Fig. 7 Tracing of a silica gel thin-layer plate showing the mobilities of authentic and putative glucose, cellobiose and cellotriose from acid hydrolysis of A compound glucan. Solvent 1-butanol--acetic acid-ethyl ether-water (9:6:3:1).

Discussion

Approximately one-half of the total IAA of mature sweet corn kernels is in ester linkage to a cellulosic glucan. Assuming one mole of IAA per mole of glucan, the chain length of the glucan would be 7 to 50 glucose residues. Glucan plus IAA account for about 50% of the weight of the A compound. The remaining 50% of the compound is as yet unidentified. On the basis of incomplete evidence the unidentified moieties appear to be polyhydric nitrogen containing compounds of mass about 300. These substituents are sufficiently lipophylic so that the A compound is maximally soluble in 70 to 90% acetone in water.

Nothing is known of the physiological role of the IAA esters. It is possible that they are storage forms of the hormone, detoxification forms, transport forms or the physiologically active hormonal compounds. Understanding the importance, if any, of these compounds will depend upon knowledge of a) their ubiquity in nature b) their presence in growing vegatative tissue where hormone concentrations are limiting* and, c) development of *in vitro* tests for hormonal activity. Bioassays for growth promoting activity may not provide the requisite information since these assays exclude all compounds which do not readily permeate a cell membrane. One might ask whether Coenzyme-A would have been discovered by rubbing a lanolin paste of Coenzyme-A on the skin of a rabbit or by immersing the rabbit in a buffered solution of the Coenzyme?

---

*Interestingly a recent report by Davies and Galston (13) suggests the formation of an IAA polysaccharide complex formed upon feeding pea stem sections with labeled IAA.

Our current data are too limited to warrant much speculation. It is a fact that all, or almost all, of the IAA of corn kernel tissue is esterified. The alcohol moiety is a cyclitol, cyclitol glycoside, a sugar or a glucan. Thus, there is a possibility that these compounds could function in transglycosylation reactions. If so the inclusion of this work in this symposium will, in fact, be warranted.

References

1. D. E. Atkinson, *Ann. Rev. Biochem.* 35, 85 (1966).
2. Z. Piskornik and R. S. Bandurski, *Plant Physiol.* 50, 176 (1972).
3. M. Ueda and R. S. Bandurski, *Plant Physiol.* 44, 1175 (1969).
4. R. H. Hamilton, R. S. Bandurski, and B. H. Grigsby, *Plant Physiol.* 36, 354 (1961).
5. G. S. Avery, J. Berger and B. Shalucha, *Amer. J. Bot.* 28, 596 (1941).
6. A. J. Haagen-Smit, W. D. Leech and W. R. Bergren, *Amer. J. Bot.* 29, 500 (1942).
7. C. Labarca, P. B. Nicholls and R. S. Bandurski, *Biochem. Biophys. Res. Commun.* 20, 641 (1965).
8. P. B. Nicholls, *Planta* 72, 258 (1967).
9. T. Takano, R. S. Bandurski and A. Kivilaan, *Plant Physiol.* 42, S-13 (1967).
10. M. Ueda, A. Ehmann and R. S. Bandurski, *Plant Physiol.* 44, 1 175 (1969).
11. A. Ehmann and R. S. Bandurski, *Plant Physiol.* 47, S-14 (1971).
12. A. Ehmann and R. S. Bandurski, *J. Chromatog,* 72, 61 (1972).
13. P. J. Davies and A. W. Galston, *Plant Physiol.* 47, 435 (1971).

FIBRILLAR, MODIFIED POLYGALACTURONIC ACID IN, ON AND BETWEEN PLANT CELL WALLS*

J. Ross Colvin

Division of Biology Sciences
National Research Council of Canada
Ottawa, Canada

and

Gary G. Leppard

Water Science Subdivision
Inland Waters Directorate
Department of the Environment
Ottawa, Canada

Abstract

Fibrillar polygalacturonic acid is found on the outside of, between and sometimes within the cell walls of suspension cultures of plant cells. It may also be found projecting from the cell walls of whole plants. The fibrils are heterogeneous in both length and width and have no internal, close-range order. They associate in extracellular spaces to form yarn-like aggregates or sheets. Proteins and other ultraviolet absorbing groups are strongly adsorbed or attached to the fibrils. It is possible that the fibrils serve as skeletal material in

* Issued as N.R.C.C. No. 12925.

plant walls, as a "filler" substance and cementing agent between cells, as a carrier of lignin precursors or as a promotor of symbiosis.

Introduction

The greater part of most biological tissues is fibrillar, either at the microscopic or submicroscopic level. For this reason there has been a continuing search for better tools or methods to permit the detection of some sort of orientation in those native substances which are considered amorphous. This search is still underway for the pectic substances which are most abundant in the primary cell walls and intercellular layers of the soft tissues of land plants and which are primarily, but not necessarily exclusively, ($\alpha 1 \rightarrow 4$) polymers of D-galacturonic acid (1). The pectic substances have been generally assumed to be lacking in orientation in native tissues although Nakamura (2) nearly forty years ago observed double refraction in cell-wall pectins of FRITILLARIA pollen mother cells. Some years later, Roelofsen and Kreger (3, 4) obtained additional evidence for oriented pectin molecules in fresh collenchyma cell-walls of the petioles of Petasites vulgaris. They reported the association of pectin molecules into fibrils in this tissue but their study was not extended to other cell walls. At present, possibly because of certain reservations about earlier work (5), it is generally assumed that pectic substances are amorphous and without characteristic form.

The purpose of the present article is to summarize the results of a series of studies on fibrillar polygalacturonic acid, to relate these results as a group to previous observation and to speculate on the significance of oriented pectic substance for plant tissue development.

Structure and Morphology

To begin with, it is certain that the recent recognition of fibrils of polygalacturonic acid on the surface of cultures cells of several species (6) was not the first time that fibrous pectin had been observed. Roelofsen and Kreger (3, 4) reported it twenty years ago; fibrillar material of uncertain composition but resembling the fibrillar aggregates of Leppard *et al*. (6) was reported by White (7) from observations with the optical microscope on cell cultures of *Picea glauca*; and Sutton-Jones and Street (8) observed electronopaque fibrils on the outside of cultured cells of *Acer*. Almost certainly, Halperin and Jensen (9) saw them as a bushy coat on the outside of cells of *Daucus* but thought they were cellulose.

The detailed form of the fibrils was first established by recent observations on cells of suspension cultures of *Daucus*, *Ipomoea* and *Phaseolus* (6). The breadth of the flat ribbons may vary from the limit of resolution of sectioned material in the electron microscope up to about 200 Å. There seems to be no definite width and the fibrils sometimes appear to be split longitudinally or branched suggesting that they may have a fasciculate structure internally. The length of individual fibrils is indefinite but some may extend up to one-half micron. From time to time, the individual fibrils may associate on the surface of cells to form fuzzy yarnlike aggregates which may be twenty or more microns long. One such bundle and its ribbonlike component fibrils is shown in Fig. 1.

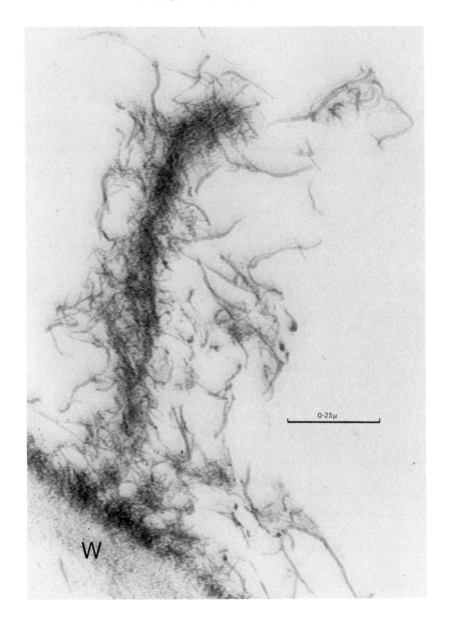

Fig. 1 Short, rope-like aggregate of fibrillar
material outside the cell wall of *Phaseolus*.
Fixation was first by 6% glutaraldehyde followed by
ruthenium-osmium staining. Note the flat ribbon-
like strands which branch or become thinner even when
they do not branch. Such strands often form a thick
coat about the cell(s). Sometimes the fuzzy, rope-
like aggregates are many microns long (6). W - wall

The longitudinal extension, the branching and the
splitting of the fibrils suggested that internally
these threads might have some detectable order. However, no conclusive evidence exists at present for
either short-range or long-range order within a
fibril (12). In addition, the fibrils may associate
between cells or on the surface of cells to form
sheets which in ordinary tissues would be said to be
a part of the middle lamella (6, 10, 11). Speaking
generally, the fibrils in suspension cultures often
seem to form a loose but definite and coherent network between the cells of aggregates of the suspension. Within plant tissues, this network may be
compressed and the details of the fibrils obliterated;
this problem has no satisfactory experimental solution at the present time.

Physical and Chemical Properties

The physical and chemical properties of these
threads set them apart from any other fibrillar
components of plant tissue. First, although they
may be observed in section with difficulty in the
electron microscope, they do not absorb strongly any
of the usual electron stains (6). They are seen most
clearly when fixation with glutaraldehyde is followed
by ruthenium-osmium staining. No other common
component of plant cell walls reacts in this way,
except for certain granules which may be a form
variant of the fibrils (11). Second, the fibril
aggregates in suspension have an ultraviolet

absorption spectrum similar to that of lignin but
the extinction coefficient is only about 1/5 as great
(12). Strong ultraviolet absorbance may also be
observed by ultraviolet-microscopy of wet mounts of
whole cells and of sections of embedded whole cells
(11, 12). Third, the infrared spectrum of isolated
fibrils is in general consistent with that of pectin
(12) but lacks resolution. Fourth, the fibril coat
on the outside of whole cells gives a positive
Wiesner reaction (method according to White (7) for
lignin and it is also soluble to some extent in
aqueous dioxane (6). Finally, the fibrils are
resistant to the action of lipase, pronase, pectinase
and trypsin under conditions where these enzymes (or
mixtures of enzymes) vigorously attack their normal
substrates (6). These properties, taken as a whole,
are not consistent with any simple interpretation
of the composition of the observed fibrils. Nevertheless, they can be explained by the assumption of
a modified polygalacturonic acid composition as
summarized in the next section.

Composition of the Fibrils

At first, the composition of the observed fibrils
was thought to resemble lignin (6) because of their
ultraviolet absorption, the positive Wiesner reaction,
their partial solubility in aqueous dioxane, the
resistance to pectinase degradation and the presence
in quantity of a structured form of Klason lignin
(6). However, subsequent analyses on *Ipomoea* samples
showed that this preliminary identification could not
be wholly correct. The first indication that the
fibrillar material was not predominantly lignin came
from determinations of its methoxyl content which was
3.2% whereas normal wood lignins vary from 13-23% (12).
In addition, water dispersions of the fibrils showed
a strong qualitative reaction with the carbazole
reagent for hexuronic acids, although no galacturonic
acid residues (or other uronic acid) were detected in

acid hydrolysates of the fibrils and they were
resistant to the action of pectinase. Subsequent
quantitative application of the carbon dioxide
evolution method gave a hexuronic acid content of
67% and quantitative application of the carbazole
technique indicated a hexuronic acid content of 68%
(12).

There was, therefore, little doubt that the
fibrils were composed largely of hexuronic acid
residues but it remained to be established which
particular acids were involved. This was done by
propylation of the polymer, followed by borohydride
reduction. [This project represents a separate
detailed investigation which is to be submitted for
publication in a specialty journal.] The results
showed only galacturonic acid as the monomer of a
long-chain polymer. No other uronic acids were found
and no neutral reducing sugars could be demonstrated
(6, 12). Likewise, no nucleic acid bases could be
detected (6). In summary, at least two-thirds of the
dry weight of the isolated fibrils is composed of
polygalacturonic acid (ignoring, for the moment, the
weight of the various cations).

It would seem that the initial inability to find
galacturonic acid residues may be attributed to almost
complete destruction of this acid under the conditions
of hydrolysis. However, the inability of an active
pectinase preparation to degrade the fibrils on the
outisde of cells is somewhat more puzzling, especially
in view of the demonstrated lack of crystallinity of
the threads (12). Although it must be, as yet,
speculative, the sum of two possible effects may
explain this surprising resistance to enzymatic
degradation. First, the fibrils absorb ultraviolet
radiation and, therefore, must contain or have
attached some molecules or groups which absorb
ultraviolet light since pure galacturonic acid
residues do not. Such molecules or groups may
prevent a close approach of the polymer chain to the

enzyme surface thus hindering hydrolysis. Second,
the fibrils contain about 5% protein (6), either
strongly adsorbed or attached covalently to the
polymer chain. Such large residues would also inhibit the hydrolysis of the polygalacturonic acid.

Attempts to identify the groups which are
responsible for the adsorption of ultraviolet
radiation have been only partially successful. The
protein(s) of course contributes to the effect but
the proportion is unknown. One ultraviolet absorbing
and fluorescent compound has been isolated from the
product of alkaline nitro-benzene hydrolysis of the
isolated fibrils (6) and has been shown to be an
aliphatic, carboxylic acid with no hydroxyl groups
but apparently at least one ether linkage. A second
such fluorescent compound was shown to be sodium
acetate, undoubtedly a degradation product of the
galacturonic acid residues. A third ultraviolet
absorbing compound in the hydrolysate was shown to
be the product of a reaction between nitrobenzene and
any of the amino-acids cystine, serine, methionine,
tyrosine, histidine, glycine or tryptophane but not
of the other common aminoacids. It was not detected
previously (12) because it is confined almost
completely to the small nitrobenzene phase of the
hydrolysate. Identity of two other fluorescent
compounds has not yet been determined.

In summary, the newly recognized fibrils found
in suspension-cultures of *Ipomoea* are mostly
polygalacturonic acid or its salts. At present, there
is no reason to doubt that this is probably true for
the fibrils of other species. The fibrils are not
pure polygalacturonic acid because, in addition to
protein(s), there may be other as yet unidentified
ultraviolet absorbing groups.

Occurrence of Fibrillar Polygalacturonic Acid

The recognition of a distinct structure or

component in what is essentially an artificial
system always raises the problem of whether the
entity occurs with, or has significance for, natural
tissues or cells. As reported previously (6), these
heterogeneous fibrils of a characteristic morphology
are found in suspension cultures of *Ipomoea, Daucus*
and *Phaseolus* (6). On the basis of the electron
micrographs by Sutton-Jones and Street (8), there
seems little reason to doubt that they also occur in
suspension cultures of *Acer*. Equally, there is every
reason to suppose that these fibrils were also
observed by Halperin and Jensen (9) as a bushy coat
outside *Daucus* cells. In addition, the optical
microscope studies by White (7) suggest that they may
occur in cultures of *Picea*. Fibrillar material can
be seen by scanning electron microscopy on the outside of suspension cultured cells of *Ipomoea* (Fig. 2)
and one must emphasize that these fibrils are not
artifacts due to dessication because bundles of them
can be observed directly in wet mounts by ultraviolet
microscopy (11, 12). Likewide, fibrillar material of
the same morphology and staining properties has
recently been observed in the external mucilage layer
of root tips of whole plants of *Ipomoea*, of *Daucus*
and of Apex wheat (*Triticum*) (Fig. 3) grown in the
laboratory. Taken as a whole, this collection of
observations is substantial evidence that the fibrils
are a constituent of many plant cell cultures and
tentative evidence for their presence in tissues of
whole plants grown in the laboratory.

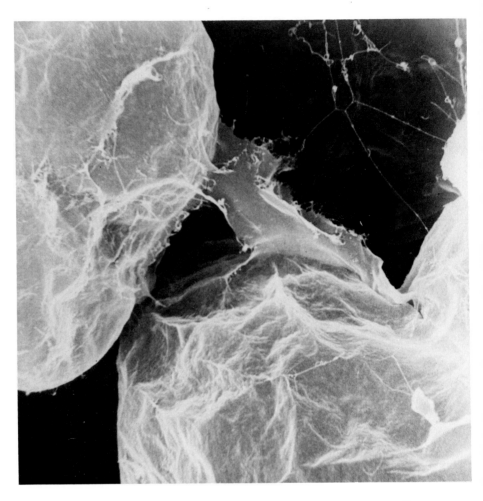

Fig. 2 Scanning electron micrograph of the surface of suspension-cultured cells of *Ipomoea*. Note the evidence for fibrils and bundles of fibrils on the surface. We wish to thank Prof. D. Brown, Dept. of Biology, University of Ottawa for his assistance with the photography.

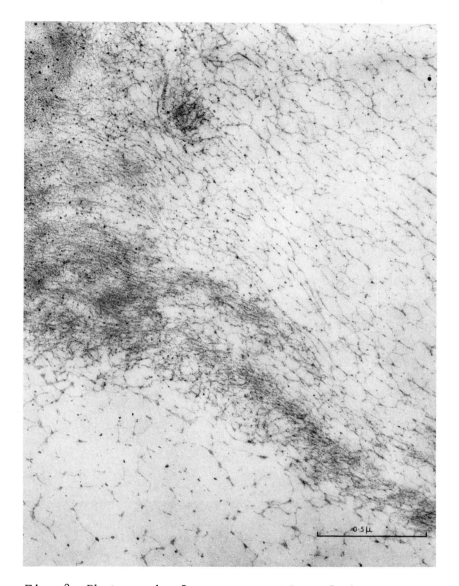

Fig. 3 Photograph of a cross-section of the external "mucilage" layer of a root-tip of *Triticum*
(variety Apex). Fixation by glutaraldehyde followed
by staining with ruthernium-osmium. Note the fibrils
which are similar in form to those on the surface
of suspension-cultured cells.

Some evidence exists that the same or similar fibrils are found within tissues of whole plants grown under natural conditions (3, 4, 11). However, this evidence is as yet too sparse to be completely convincing as to the widespread occurrence of these fibrils in native tissues. Because the substance in and between the cell walls of most plant tissues is tightly compressed, it would be difficult to observe any fibrils even if they were present. Cell dispersion in suspension cultures is the principal reason why the fibrils can be detected there. In the same way, the relatively uncompressed external surfaces of root cells permit their observation. For this reason, direct observation of such fibrils in whole plants will likely remain the exception, not the rule.

At the present time, all work on the composition of the fibrils has been limited to those from *Ipomoea*. There is, therefore, no direct evidence that the extracellular fibrils from other species which stain with ruthenium-osmium are chiefly composed of polygalacturonic acid. More work is necessary. However, at present it is reasonable to invoke Occam's razor and to assume, with no facts to the contrary, that the general composition of the fibrils from other species is the same as, or similar to, those from *Ipomoea*.

Biogenesis of Fibrillar Polygalacturonic Acid

Clearly, in principle, the biogenesis of these fibrils can be separated into two phases; (1) the elaboration or biosynthesis of the monomer(s) and (2) their transport and physical incorporation into threads. At present, almost nothing is known definitely about either phase. With the knowledge that the fibrils are chiefly polygalacturonic acid, it is reasonable to suppose that the chief precursor is *myo*-inositol (or a compound which shares a pathway

with *myo*-inositol) as in other plants (13, 14). No
definitive work has yet been done on the fibrils *per
se*.

Likewise, no definitive data are available on
the site of synthesis of the polygalacturonate(s),
its means of transport (if indeed it is transported)
or the site of its incorporation into the fibrils.
If one assumes that a portion of the "slime droplet"
material which accumulates outside the roots of *Zea*
and which contains some galacturonic acid also contains this fibrillar substance, it could be argued
that the site of polymerization is the Golgi apparatus
(15) with transport across the membrane by vesicles.
This, however, is pure speculation.

Another set of problems about which nothing is
known is the mechanism of linear extension of the
individual polygalacturonic acid chains and their
association to form extended aggregates. It might
be reasonable to suppose that the linear extension
of individual chains is due to Coulomb repulsion of
the ionized carboxyls along the polymer but if this
were so then it also implies repulsion between
adjacent chains thereby hindering consolidation into
observable aggregates. On the other hand, if
repulsion between carboxyls is minimized by either
esterification or by a shield of ions, it is difficult to understand why the polymer chain would not
tend to take up a random coil arrangement.

Significance of Fibrillar Polygalacturonic
Acid for Plant Tissues

As outlined in the foregoing sections,
considerable evidence now exists for fibrillar,
polygalacturonic acid in and between the cell walls of
plant cells in culture and in some whole plant tissues.
It is reasonable and natural to ask what role this
substance has in development of the cell or the plant.

Because it is a fibrillar polymer, polygalacturonic
acid might act as a structural component but this
seems unlikely. The polymer is unconsolidated
physically, seems to be weak elastically and is
unlikely to be much use structurally except as a
"filler" substance. However, this function combined
with a cementing action, both within the plant cell
wall and between plant cells, may well be important.
At the moment, it is not possible to estimate what
contribution this component makes to cell wall
rigidity or to the stiffness of tissues. However,
it may be considerable and this possibility becomes
more interesting in view of Northcote's suggestion
(16) that a polysaccharide coated with lignin
precursors and located between cells may be involved
in lignification.

Although the suggestions was made from another
point of view, Halperin and Jensen (9) have indicated
that fibrillar material on the cell surface, which
they presumed to be cellulose, might be involved in
the process of cell separation. They noticed that
the fibrils grow profusely until older walls become
thick and bushy and that the fibrils were sloughed
into the surrounding medium in great quantities. The
role of these effects in tissue development remains
to be investigated.

Depending upon the definitions used, this fibrillar material may be regarded as a secretion product.
Such secretion products may be important in some cases
of symbiosis. Dart and Mercer (17) and later Dart
(18) have shown that a mucilaginous layer on and
outside of legume roots participates in the development of colonies of *Rhizobium* about the roots. Their
photographs of the layer show that, in cross-section,
it has a granular component similar in dimensions to
cross-sections of the fibrillar tufts seen in sections
of cell cultures. Furthermore, photographs of metal-
shadowed surfaces of the roots show fibrils which are
similar in dimensions to those seen in cell cultures.

It is therefore possible that this so-called mucilaginous layer has a composition similar to that seen in cell cultures. It is an appealing idea that an extracellular material, which may be polygalacturonic acid, may serve to anchor helpful foreign cells as well as constitutive ones.

It should not be forgotten that such mucilaginous (gel-like?) layers may serve to localize enzymes or enzyme complexes as well as cells. The layer observed by Dart and Mercer (17) and by Dart (18) is known to be associated with a polygalacturonase (17), apparently specifically induced in the host. Halperin (19) has also shown that acid phosphatase may be found in the same region as the fibrillar material between cells. Whether the association is by covalent bonding or by physical adsorption is not yet clear. Localization by simple entrapment of large molecules or complexes is also possible.

The demonstrated presence of fibrillar polygalacturonic acid in, on and between plant cell walls is a warning that, henceforth, cellulose cannot be regarded as the sole fibrillar component of plant cell walls. In the past, it has been generally assumed that threads of the appropriate size were cellulose microfibrils in the wall or were material outlined by such microfibrils. In the future more care will need to be taken in the identification of plant wall fibrils.

Acknowledgments

The authors gratefully acknowledge the help of Mr. R. Whitehead with the photographs.

References

1. G. O. Aspinall, *Polysaccharides*. Oxford, Pergamon Press (1970).
2. T. Nakamura, *Cytologia*, p. 482 (1937).
3. P. A. Roelofsen and D. R. Kreger, *J. Exp. Bot.* 2, 332 (1951).
4. P. A. Roelofsen and D. R. Kreger, *J. Exp. Bot.* 5, 24 (1954).
5. R. D. Preston, *J. Exp. Bot.* 3, 437 (1952).
6. G. G. Leppard, J. R. Colvin, D. Rose and S. M. Martin, *J. Cell Biol.* 50, 63 (1971).
7. P. R. White, *Amer. J. Bot.* 54, 334 (1967).
8. B. Sutton-Jones and H. E. Street, *J. Exp. Bot.* 19, 114 (1968).
9. W. Halperin and W. A. Jensen, *J. Ultrastructural Res.* 18, 428 (1967).
10. G. G. Leppard and J. R. Colvin, *J. Cell Biol.* 50, 237 (1971).
11. G. G. Leppard and J. R. Colvin, *J. Cell Biol.* 53, 695 (1972).
12. G. G. Leppard and J. R. Colvin, *J. Polymer Sci.:* Part C, No. 36, pp. 321. (1971).
13. R. M. Roberts, J. Deshusses and F. Loewus, *Plant Physiol.* 43, 979 (1968).
14. F. Loewus, *Annu. Rev. Plant Physiol.* 22, 337 (1971).
15. D. D. Jones and D. J. Morre, *Z. Pflanzenphysiol.* 56, 166 (1967).
16. D. H. Northcote, *Essays in Biochemistry* 5, 89 (1969).
17. P. J. Dart and F. V. Mercer, *Archiv Mikrobiol.* 47, 344 (1964).
18. P. J. Dart, *J. Exp. Bot.* 22, 163 (1971).
19. W. Halperin, *Planta* 88, 91 (1969).

CELL SURFACE POLYSACCHARIDES OF THE
RED ALGA *PORPHYRIDIUM*

J. Ramus

Department of Biology, Yale University
New Haven, Connecticut   06520

Abstract

The unicellular red alga *Porphyridium aerugineum*
is encapsulated by gel state polysaccharides. The
polysaccharides are composed primarily of glucose,
galactose, xylose and ester sulfate, and several minor
components including hexuronic acids. They are highly
acidic, form ionic bridges through divalent cations,
and in this form have an especially high molecular
weight.

Dimensions of the polysaccharide capsule vary,
being thinnest in log phase cells, and thickest in
stationary phase cells. Using ^{14}C-labeling techniques,
the rate of solubilization from the cell surface as a
function of the rate of deposition was calculated.
The rate of deposition exceeds the rate of solubilization in stationary phase, allowing a capsule to form.

When incubated in the presence of $^{35}SO_4^=$,
Porphyridium cells remove the isotope from the medium
and incorporate it rapidly into the extracellular
polysaccharide as ester sulfate.

Preliminary electron microscope evidence suggests
that the synthesis, movement and deposition of the

333

polysaccharides on the cell surface is mediated by
the Golgi apparatus and derived vesicles.

Symbols

CPC = cetyl pyridinium chloride
CPM = counts per minute
EDTA = ethylenediaminetetracetate
GDP = guanosine diphosphate
O.D. = optical density
PAS = periodic acid-Schiff's reaction
TLC = thin layer chromatography
Tris = tris(hydroxymethyl)aminomethane
UDP = uridine diphosphate
V_e = elution volume
V_o = void volume
V_t = total volume

Introduction

Cells of the red algae are always generously
coated with "gel state" (1) polysaccharides. Structural biochemists found these polysaccharides to be
unique, and designated for them the family "sulfated
D- and L-galactans" (see 2, for review), or the
"agar" family (1). Within the family there is great
structural variation, i.e., the group includes mostly
heteropolysaccharides.

These cell surface polysaccharides exist in a
characteristic gel state (1), i.e., "a fairly per-
manent network structure formed from polymer solutions".
Typically, the interconnected network gives rise to
the characteristic texture and properties of the
polysaccharides. In the interstices of the network
are molecules of solvent and solute. The level of
structural organization varies from cyrstal lattice,
to fibrous, to completely amorphous, depending on the
degree of cross-linking between polymers. Their
solubility in water depends both on the chemical nature
and number of bonds between polymers, but most tend to

be water soluble.

The biological function of these polysaccharides
is at present the subject of learned speculation only.
Many are acidic, a property conferred by the carboxyl and ester sulfate functional groups of constituent sugars. There is evidence (3, and some to
be presented in this paper) that these acidic polymers possess a cation retention and exchange capacity, making them likely ion reservoirs contigous
with the cell membrane. The gel state characteristics
confer a hygroscopic nature on the polysaccharides,
which may maintain hydration of red algae cells during
periods of desiccation (2), i.e., low tide, low
water, or drought. Further, all thalloid red algae
are either aggregates of unicells, or aggregates
of distinct filaments, each of which is derived from
a single apical cell. It seems likely that gel
state polysaccharides serve as an intercellular
(interfilament) cementing material, a binding matrix,
allowing pseudoparenchymatous tissues to be built
from ontogenetically distinct structural units.

The state of knowledge concerning the biological
function, synthetic pathways, and means of deposition
of red algal cell surface polysaccharides is at
present in its infancy. There is then a biological
imperative to study the production of these seemingly
important macromolecules, both to understand their
role in the red algae, and, in general, to understand the properties of cell sufaces conferred by
extracellular polysaccharides.

Porphyridium aerugineum is a unicell, and is
morphologically the simplest of the red algae. The
cell is encapsulated by a thick layer of water
soluble polysaccharide. In liquid medium, the polysaccharide becomes available in its native state,
making rigorous extraction procedures unnessary.
Pure cultures attain densities of 25 X 10^6 cells/cc

on defined medium in short periods of time, and release large quantities of polysaccharide. Cultures
represent a homogeneous tissue producing a single
polysaccharide, unlike other cell suspension systems
which may undergo differentiation. In addition,
much is known of the physiology of the species (4,
5, 6). *Porphyridium aerugineum* seems, therefore,
to be ideal for the study of the biogenesis of
cell surface (extracellular) polysaccharides.

The mission of this paper then is to review
the status of our knowledge concerning the production of capsular polysaccharide by *Porphyridium
aerugineum*, to present some new data concerning both
polysaccharide structure and kinetics of production,
and compare this data with that obtained from experiments with other red algae.

Polysaccharide Isolation and Characterization

We have confined our studies to a single species
of *Porphyridium*, *P. aerugineum*, isolated from soil.
The cells are blue-green in color (rather than red),
due to the presence of chlorophyll and phycocyanin
in the chloroplasts, and the absence of the usual
masking phycoerythrin (7). The cell structure has
been described in detail elsewhere (7, 8). This
red alga is grown routinely in a defined low salts
liquid medium, tricine-buffered at pH 7.6, 25°C, low
light intensity and the cultures agitated by air, air
supplemented with 5% CO_2, or gyrorotary shaking (8).

As cells grow in liquid medium, the medium
rapidly increases in viscosity due to the slow
liberation (solubilization) of capsular polysaccharide
from the cell surfaces. The polysaccharide can be
removed quantitively (8) from cell-free culture
medium (of low ionic concentration) by (1) precipitating with the cationic detergent CPC, (2) washing
the water-insoluble precipitate to remove excess
detergent ion, 3) resolubilization of the precipitate

in 2M Ca^{++} which exchanges for the CPC on the polymer, and the 4) reprecipitation of the pure Ca^{++} salt
of the polysaccharide in ethanol. Any number of
monovalent or divalent cations can be used to dis-
sociate the CPC from the polysaccharide, but Ca^{++} is
used preferentially because of its high solubility
in ethanol, and the high solubility of the Ca-
-polysaccharide salt in water.

Subsequent qualitative comparison of solubilized
polysaccharide isolated from culture medium with
capsular polysaccharide extracted (hot water) from
cells showed the two fractions to be identical (8).
Therefore, it is capsular polysaccharide that is
solubilizing into the medium, and not polysaccharide
from some unknown cellular source. Quantitative
precipation by the cationic detergent CPC indicates
that the polysaccharide is highly acidic (anionic).
Cationic dyes, as alcian blue 8GX (derived from
copper phthalocyanin), can also be used for precipitation, resulting in a colored product (9).

Constituent sugar identification was made from
the hydrochloric or trifluoroacetic acid hydrolysates
of the polysaccharide (8, 9) on silica gel thin
layer chromotographs. The bulk of the polymer is
made of xylose, glucose and galactose, with four
other constituents. Subsequent gas chromatography
(10) of trifluoroacetic acid hydrolysates corroborated
the presence of mostly xylose, glucose and galactose,
and in addition tentatively identified small quantities of guluronic acid, galacturonic acid, another
pentose, and an anhydrohexose. The barium chloran-
ilate assay revealed the polysaccharide to be 7.6%
sulfate by weight, although to which sugar the
sulfate is bound is presently unknown. The presence of ester sulfates and uronic acids in the
polymer account for its acidic properties, and hence,
its CPC-precipitability. Attempts to precipitate
the polysaccharide in the presence of Ba^{++}, Ca^{++} and
Mg^{++} (as Cl^-, OH^- or $SO_4^=$) failed.

337

The capsular polysaccharide of *Porphyridium* is therefore, a complex heteropolymer not unlike the sulfated galactans of other red algae(2). However, is the polysaccharide homo- or heterogeneous, i.e., does the capsule contain a single (qualitatively) heteropolymer, or several kinds of heteropolymers? The question was partially answered by fractionating both ^{14}C- and ^{35}S-labeled polysaccharide by gel electrophoresis and gel filtration. Isotopically labeled polysaccharide was produced by incubating cells in the presence of $H^{14}CO_3^-$ or $^{35}SO_4^=$, then isolating the excreted polysaccharide from cell-free medium. The material was migrated in an electrophoretic field on weak (3%) polyacrylamide gels in a pH 8.3 Tris-EDTA-borate buffer (9). The polysaccharide was localized on the gels after migration by PAS or alcian blue staining. Radioactivity in gel slices was determined by liquid scintillation counting. Electrophoresis divides the capsular polysaccharide into two fractions, one which migrates into the gel and contains little sulfate, the other which remains at the origin and is sulfate rich.

Fractionation of the polysaccharide was attemped by gel filtration. In all cases, a pH 7.6 Tris(.01M)-EDTA(.01M)-NaCl(.05M)-sodium azide (.02%) buffer, a K15/90 column (Pharmacia Fine Chemicals), and ^{14}C- or ^{35}S-labeled polysaccharide were used. Approximately 500 µg of polysaccharide in 1 ml buffer was loaded onto the column, the flow rate maintained at 10 ml per hr, and 8 ml fractions collected. The elution profile of labeled polysaccharide was monitored as the CPC-precipitable material in the column eluate. On Bio-Gel A50m (Bio-Rad Laboratories, exlusion limit 50 X 10^6 daltons), the polysaccharide eluted in the void volume (V_o) as determined by Blue Dextran 2000 (Pharmacia Fine Chemicals). It was difficult to believe that the capsular polysaccharide exceeded $\overline{MW}$ 50 X 10^6. An

alternative explanation was that the polysaccharide
(being a gel state polymer) was forming cross-
-linkages, perhaps through anionic functional groups
(carboxyl and ester sulfate) with divalent cations,
resulting in a molecule of infinite molecular weight.
In that the polysaccharide is rich in ester sulfate
and low in uronides, it is likely that the ester
sulfates are responsible for extensive cross-linking
(simple ionic bridging). To test this hypothesis,
both ^{14}C- and ^{35}S-labeled polysaccharides were subjected to mild acid hydrolysis, mild enough to
hydrolyze the sulfate esters, but not glycosidic
linkages. In the absence of sulfate ester, cross-
-linkages should be at a minimum, and therefore the
polysaccharides, now of lower (finite molecular
weight, should be retained on molecular sieve columns.
Indeed, this was the case, as seen in Fig. 1.

A Sephadex G-200 column was calibrated (Fig. 1)
with 5 calibration solutes, Blue Dextran (MW 2 X 10^6,
OD$_{620}$), Dextran T150 (MW 1.5 X 10^5, phenol-sulfuric
acid assay, OD$_{490}$), hemoglobin (MW 6.4 X 10^4, OD$_{541}$),
Vitamin B$_{12}$ (MW 1.35 X 10^3, OD$_{510}$) and Na$_2$Cr$_2$O$_7$
(MW 2.62 X 10^2, OD$_{390}$).

Polysaccharide directly from cell-free medium
or CPC-precipated polysaccharide gave the same
elution profiles, indicating the CPC isolation
procedure does not alter the physical conformation
of the polysaccharide, as noted previously by other
means (8, 9). The unaltered capsular polysaccharide
is excluded from the Sephadex G-200 column (Fig. 1).
Mild acid hydrolysis (0.01M HCl, 100°C for 30 min),
however, radically alters its elution profile. Mild
acid hydrolysis does not release sugar monomers, as
monitored by silica gel TLC; it does, however,
release free sulfate (Fig. 1, ^{35}S hy-p) as monitored
by liquid scintillation counting (Inter-technique

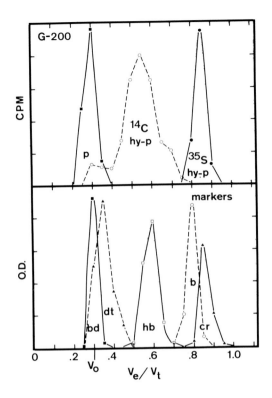

Fig. 1 Fractionation of capsular polysaccharide
(p) and its weak acid hydrolysates (hy-p) on a
Sephadex G-200 column; 1.5 X 90 cm; flow rate 10 ml/hr;
pH 7.6 0.01M Tris-EDTA buffer, 0.05M NaCl, 0.02%
NaN$_3$; 8 ml fractions; reference solutes (markers)
used were Blue Dextran 2000 (bd), Dextran T150 (dt),
hemoglobin (hb), vitamin B$_{12}$ (b) and Na$_2$Cr$_2$O$_7$ (cr).

SL30; Aquasol-Nuclear Chicago). Further, after mild
acid hydrolysis, no CPC-precipitable ^{35}S activity is
recovered, indicating hydrolysis of the ester sulfate was complete. Mild acid hydrolyzed ^{14}C-labeled
polysaccharide (^{14}C hy-p) results in CPC-precipitate
activity being retained by the column. Even though
the polysaccharide has been de-sulfated, it is still
sufficiently anionic via the carboxyl groups of the
constituent uronides to allow precipitation by CPC.
The resulting elution profile (Fig. 1) is very
broad, indicating that the degree of cross-linking
has been reduced, and that the polysaccharide is now
a mixture of smaller units, ranging in $\overline{MW}$ from 1.3 X
10^3 to 2 X 10^5, with most pieces ranging from $\overline{MW}$
6.4 X 10^4 to approximately 1 X 10^5. The possibility
of oligosaccharides being formed as a consequence
of the hydrolysis of some especially labile glycosidic
linkages cannot, unfortunately, be ruled out and thus
probably account for some of the lower molecular
weight pieces. However, from the electrophoretic
and molecular sieve data, it is concluded that the
polysaccharide is heterogeneous.

The polysaccharide in its native (cross-linked
or gel) state should be massive, and consequently,
visible in the electron microscope. Therefore,
pure polysaccharide in solution was spread on a
formvar-carbon coated grid, negative-stained (2%
potassium phosphotungstate for 1 min) and observed.
This technique shows the polysaccharide (Fig. 2) to
be massive, forming a 3-dimensional network which
perhaps accounts for its behavior on molecular sieve
column.

The agarose fraction of agar isolated from the
red seaweeds *Gracilaria* and *Gelidium* has a great tendency to gel, a function of the agarobiose repeating
structure in the agarose fraction (1). The polysaccharide of *Gloiopeltis furcata,* which differs
from agarose in containing a high proportion of

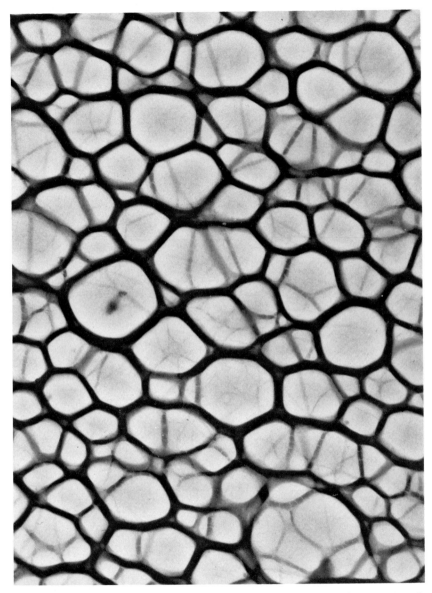

Fig. 2   Electron micrograph of a negatively stained
(positive print) pure capsular polysaccharide solution; 0.5 mg/ml Ca-polysaccharide; 1 min stain with
2% potassium phosphotungstate, pH 7.0; X40,000.

ester sulfate, forms only a viscous solution, but not a stiff gel. In this regard, the capsular polysaccharide of *Porphyridium aerugineum* is most like that of *Gloiopeltis*.

We are far short of a molecular formula for the capsular polysaccharide of *Porphyridium aerugineum*, but some progress had been made in its characterization. In the future, we hope to determine the molar ratios of constituent monomers, linkages, location of ester sulfates and number of heterogeneous types.

Kinetics of Capsular Polysaccharide Production

The anthrone (12) or phenol-sulfuric acid (13) assays were used to monitor solubilized polysaccharide in cell-free medium, and cell density was determined by a Coulter electronic particle counter or hemacytometer. Cells in gyrorotary shaken cultures attain a maximum density of 5×10^6 cells/cc (Fig. 3), whereas cultures bubbled with an air-5% CO_2 mixture attain maximum densities of 25×10^6 cells/cc (Fig. 4) in the equivalent time. CO_2 becomes limiting for gyrorotary shaken cultures at densities of 5×10^6 cells/cc, and under these conditions cells will undergo no further division, even if transferred to fresh medium. Thus, cells can be forced into stationary phase by limiting the CO_2 concentration, a useful characteristic as shall be discussed later.

The rate at which capsular polysaccharide accumulates in liquid medium, and hence the rate of solubilization, varies with the growth phase of the culture (Figs. 3, 4). In the lag phase, the rate of solubilization declines; in exponential or log phase, the rate reamins constant; in stationary phase, the rate increases. Why?

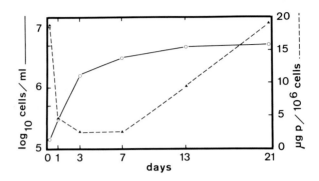

Fig. 3.   Accumulation of excreted polysaccharide
(μg/10^6cells) during lag, log and stationary growth
phases ($\log_{10}$ cells/ml) in gyrorotary shaken cultures.

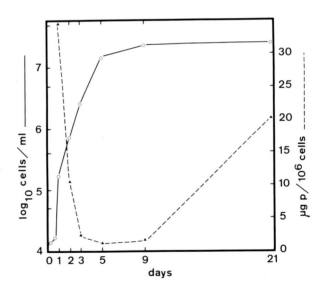

Fig. 4.   Accumulation of excreted polysaccharide (μg/
10^6 cells) during lag, log stationary growth phases
($\log_{10}$ cells/ml in cultures agitated with an air-5%
CO_2 mixture.

344

Assuming the solubility of the polysaccharide remains constant throughout the growth phase, then the rate of solubilization must be a function of the rate of deposition on the cell surface. Simply stated, the more capsular polysaccharide deposited, the more solubilizes; the less capsular polysaccharide deposited, the less solubilizes. A series of experiments were conducted to test this hypothesis.

Pulse-chase experiments were used for rate calculations. Encapsulated cells were uniformly labeled with ^{14}C (1 μCi/ml $H^{14}CO_3^-$, 36 hr), thoroughly washed, resuspended in fresh (cold) medium at a high enough density so that no further division will occur, i.e., "forced" stationary phase. Here, the rate of solubilization should be highest, and linear (Fig. 3, 4). The cells were divided into 3 equal aliquots, one grown in continuous light, one grown in continuous darkness, and one fixed (killed) with neutralized glutaraldehyde (to a final concentration of 0.5%).

The glutaraldehyde-fixed cells should produce no new polysaccharide since they are dead. By monitoring the appearance of polysaccharide in the medium, the rate of solubilization in the absence of deposition can be calculated. The light and dark grown cells should deposit new polysaccharide on their surfaces at different rates, therefore, the effect of deposition on solubilization can be calculated. The relative rates of solubilization can be determined, provided 1) that the rate of appearance of polysaccharide in the medium is linear, and 2) that the specific activity of the solubilized polysaccharide remains constant.

The quantity of polysaccharide in the medium was measured by the phenol-sulfuric acid assay, while the radioactivity of the polysaccharide was measured by 1) precipitation in an excess of CPC

(final conc. 0.01%), 2) catching the precipitate on
a GSWP Millipore filter (0.22µ pore size), 3) counting disintegrations by liquid scintillation methods
(8, 9). One difficulty arose in these experiments -
the glutaraldehyde fixative interfered with both
the anthone and phenol-sulfuric acid assays, therefore direct determinations of polysaccharide in
fixed cultures were impossible. However, as shall
be seen, indirect measurements can be based on
levels of radioactivity.

The rate of solubilization of polysaccharide in
the medium from stationary phase cells remains
constant for both light and dark grown cultures
(Fig. 5). In addition, during the course of this
experiment (52 hr), the specific activity of the
solubilized polysaccharide in both light and dark
grown cells reamins constant and equal to each other
(Fig. 6). Therefore, the two requisites stated
above for the experiment were fulfilled.

The rate of polysaccharide solubilization for
light and dark grown cells differs (Fig. 5), a fact
noted previously (8). In this experiment, the
rate of polysaccharide solubilization in the light
was 0.26 µg polysaccharide/10^6 cells/hr, while in the
dark it was 0.16 µg polysaccharide/10^6 cells/hr, or
roughly two times as great.

In that specific activity of excreted polysaccharide remained constant for both light and
dark grown cultures (Fig. 6), it is assumed that it
did also for glutaraldehyde-fixed cells. Therefore,
recovered radioactivity in the CPC precipitate can
be taken as a direct indication of the relative
quantities of polysaccharide solubilized. Indeed,
CPMs remained linear throughout the course of the
experiment for both light grown and fixed cells
(Fig. 7). The rate of appearance of ^{14}C-labeled
polysaccharide in light grown cultures is 717
CPM/10^6 cells/hr, whereas in fixed cells it was

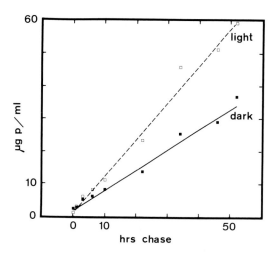

Fig. 5 Accumulation of excreted polysaccharide (μg/ml) in stationary phase of growth in continuous light and continuous darkness. Cell densities remained constant at 4.5×10^6 cells/ml.

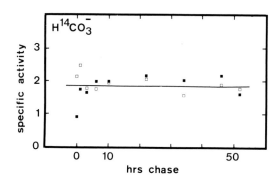

Fig. 6   Specific activity CPM $\times 10^{-3}$/μg polysaccharide) of excreted polysaccharide after a 36 hr pulse ^{14}C (1 μCi/ml NaH^{14}CO$_3$) with stationary phase cells in continuous light (open squares) and darkness (solid squares).

210 CPM/10^6 cells/hr. It would appear then that the rate of polysaccharide solubilization in the light is three times that of fixed (dead) cells. The difference can be attributed only to active deposition. It is concluded, therefore, that the rate of polysaccharide solubilization from the cell surface is a function of the rate of polysaccharide deposition at the cell surface.

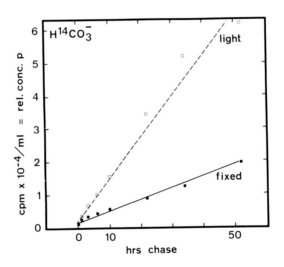

Fig. 7   Accumulation of excreted polysaccharide (CPM/ml) after a 36 hr pulse ^{14}C (1 µCi/ml NaH^{14}CO$_3$) by stationary phase cells in continuous light. Living cells (light) and dead (glutaraldehyde fixed) cells.

The capsules of cells in lag and early log phase decrease in size, in log phase they remain constant, and in stationary phase the dimensions increase.

From the experiments above it can be interpreted that the relative thickness of the capsule is a function of the relative rates of two processes, polysaccharide deposition at the cell surface, and solubilization from the cell surface. In lag and log phase, the rate of deposition is less than the rate of solubilization, thus the capsule thins. In log phase, the rate of deposition equals the rate of solubilization, thus the capsule remains thin but unchanged. In stationary phase, the rate of deposition exceeds the rate of solubilization, and the capsule increases in dimension.

Cells in log phase, especially if log phase inoculum was used, have little or no visible capsule. These cells are extremely fragile, and the integrity of the cell membrane in dependent upon Mg^{++} ion concentration (11). However, cells in stationary phase have capsules of varying thickness (up to 1μ), depending on the length of time in stationary phase, and are very resistant to physical damage.

The precise proportionality between these two parameters, i.e., the rates of deposition and solubilization, must necessarily shift as a function of still a third factor - surface area of the cell. If the solubility of the polysaccharide remains constant, then the rate of solubilization must also be directly proportional to the surface area of the capsule. The surface area, in turn, must be indirectly proportional to the rate of deposition. Capsule thickness rarely exceeds 1μ. At this point, the rate of deposition must exactly balance the rate of solubilization, which is imposed by the now

349

critical value of the surface area. In mature cells
of all red algae, the thickness of encapsulating
polysaccharide is quite uniform. It would appear
then that the thickness of capsules is regulated by
three factors, 1) the rate of deposition of the
polysaccharide, 2) its solubility, and 3) the surface area of the cell. Further, capsule thickness
must reach a maximum value as dictated by integration of these three factors.

Sulfation of the Capsular Polysaccharide

Sulfation of the capsular polysaccharide of
Porphyridium is a very rapid process (9). From the
time $^{35}SO_4^=$ is introduced into the medium, it
takes only 15 min for it to be fixed into the capsular polysaccharide, a phenomenal rate when the
complexity of the process is considered. The sulfate presumably must be taken into the cell, activated, added to sugar monomers or polymerized
sugars, carried through the cell, and deposited on
the cell surface.

Before rapid pulse-label experiments were attempted, the procedures for monitoring the appearance of label in extracellular polysaccharide
were analyzed for reliability. Routinely, the polysaccharide from measured aliquots of the medium
was precipitated by CPC, the precipitate caught
on a GSWP Millipore filter (0.22μ pore size), the
precipitate washed, dried, and the radioactivity
determined by liquid scintillation techniques. This
monitoring procedure, however, is capable of giving
quite erroneous data, especially with $^{35}SO_4^=$, and
must be subjected to control procedures. The
membrane filters themselves have a tendency to re-
tain free sulfate and other anions, a fact noted by
other workers (13). Further, the CPC also has an
affinity for free sulfate ion, and the CPC-sulfate
ion complex is retained quantitatively by the

membrane filters (11). However, both of these
isotope retention effects can be neutralized
completely by using carrier sulfate in all the
isolation procedures (11). In fact, the level of
carrier sulfate in the normal growth medium (0.4M
$MgSO_4$) is sufficient to negate the non-specific
isotope retention phenomena.

The CPC isolation protocol has been tested for
quantitative recovery of the entire capsular polysaccharide (8) using the anthrone method, and was
proved to be 97% effective. Further, using poly-
saccharide-^{35}S, equal counts are recovered from the
CPC isolation protocol, and from precipitation of
the polysaccharide in one vol. EtOH in the presence
of Ca^{++}ion. Addition of large quantities of $^{35}SO_4^=$
to cell free medium containing polysaccharide-^{35}S
does not add spurious counts to either of these
isolation procedures.

That the $^{35}SO_4^=$ is firmly bound to polysaccharide has also been thoroughly tested. Neither gel
electrophoresis (9) nor gel filtration (see section
on characterization) will separate the label from
the polysaccharide. Weak acid hydrolysis, however,
will liberate ^{35}S, indicating that sulfur is present in the molecule as sulfate ester (9).

Before rapid pulse-label experiments could be
attempted, another difficulty had to be overcome.
For the highest experimental resolution, the
labeled polysaccharide must be available immediately
for the solubilization the instant it is deposited
on the cell surface. Cells produce capsular polysaccharide at the greatest rate in late log or
stationary phase, and slowly build up a capsule.
If stationary phase cells are used, the newly
deposited polysaccharide will not appear in the
medium until overlying older polysaccharide is

351

solubilized. It is assumed here that the polysaccharide is deposited at the cell membrane and
solubilizes slowly away from the cell, forming a
continuous concentration gradient which is highest
at the cell membrane. This conclusion is supported by
observations with the election microscope (8). The
ideal situation is to use thinly-capsuled cells, or
in this case, log phase cells, However, log phase
cells produce polysaccharide at a low rate. The
manipulation required is to pulse (1 μCi/ml Na$_2$^{35}SO$_4$,
3-5 min) log phase cells in the absence of cold
carrier sulfate (9), wash the cells free of contaminating ^{35}SO$_4$$^=$, then resuspend them in complete
medium at densities at which they will no longer
divide (5 X 10^6 cells/cc for gyrorotary shaken cells),
thus forcing them into polysaccharide-producing
stationary phase. Thus an almost naked cell will
deposit polysaccharide, which becomes immediately
available for solubilization and monitoring by the
CPC isolation procedure. Under these circumstances,
^{35}S-labeled polysaccharide appears in the medium 15
min after the pulse (9).

It was found that, during short pulses, the
maximum amount of ^{35}SO$_4$$^=$ gets into sulfate-starved
log phase cells (9). However, these cells are very
fragile, and the integrity of the cell membrane must
be maintained by properly adjusting Mg^{++} levels (11).

In pulse-label experiments using log phase cells
forced into stationary phase, the specific activity
of the excreted polysaccharide should increase, then
decrease. This can be assumed if there is a free
sulfate pool in the cell, and that a significant portion of the contents of that pool is used directly to
sulfate the polysaccharide. It is impossible to
test whether the specific activity of the excreted

polysaccharide changes during a very short chase, because the sensitivity of the anthrone (5 µg/ml) and
phenol-sulfuric acid (1 µg/ml) assays simply will
not allow the measurement of the low quantities of
polysaccharide excreted immediately after the pulse.
The pulse, however, can be lengthened, as can the
period of chase, to allow accumulation of measurable
amounts of polysaccharide in the medium. Under these
conditions (4 hr pulse 1 µCi/ml $Na_2{}^{35}SO_4$, 140 hr
chase, 5 X 10^6 cells/cc), the specific activity can
be monitored (Fig. 8). During the course of the
experiment, the rate of polysaccharide excretion remains constant (Fig. 8). However, the specific activity of the polysaccharide rapidly increases, then
slowly decreases as the isotope in the pool is
diluted (Fig. 8).

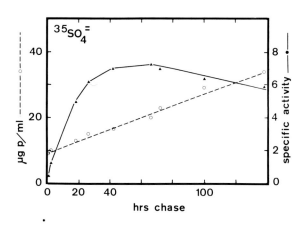

Fig. 8   Accumulation (µg p/ml) and specific activity
(CPM X 10^{-3}/µg p) of excreted polysaccharide after
30 min pulse ^{35}S (1 µCi/ml Na_2SO_4) using log phase
cells forced into stationary phase after the pulse.

Thus far, we have used *Porphyridium* only as a sphere which processes labeled isotopes. In the future, we hope to learn something of what goes on inside *Porphyridium*, i.e. the biochemical pathway leading to the sulfation of the polysaccharide.

Physical Deposition of the Cell Surface Polysaccharides

There is now growing evidence suggesting that the appearance of polysaccharides on the surfaces of red algal cells is a secretory process, and that the process is mediated by the Golgi complex. The evidence comes from two sources, 1) the experimental demonstration that the Golgi and its derivatives are the loci of synthesis and vehicles of transport of secreted macromolecules, and 2) the observation that red algal cells in the process of extracellular polysaccharide synthesis have the sub-cellular anatomy of secreting cells.

The definitive work of Jamieson and Palade (15, for bibliography) on the synthesis, intracellular transport, and discharge of secretory proteins (zymogen granules) by carbamyl choline stimulated guinea pig pancreatic exocrine cells serves as a model for experimental demonstration of the secretory role of the Golgi. As stated by Northcote (16), "the Golgi apparatus occupies a key position in the transport system of the cell, acting like a valve controlling the distribution of cellular material". He cites as examples for the discharge of polysaccharides the formation of matrix material of animal connective tissue, mucus excretion from the goblet cells in rat colon, scale production in golden (Chrysophyta) and green (Chlorophyta) algae, and cell matrix formation in higher plants. The amorphous water soluble polysaccharides of higher plant cell walls (pectin and hemicelluloses) are polymerized from constituent monosaccharides within the Golgi complex and derivative vesicles. The membrane-bound polysaccharides are transported through the cytoplasm and across the cell membrane, and are deposited on the cell surface. A proposed physical flow scheme for polysaccharide biogenesis begins at the endoplasmic reticulum, on to vesicles, the Golgi, vesicles again, and finally through the cell membrane (17). Synthetase particles isolated from higher plants catalyze the synthesis of β-glucan *in vitro* (18), albeit at very low levels. Further, the particles have been shown to be membrane-bound and Golgi--related (18), thus relating these particles to their original intracellular site (19).

Red algal cells, which are in the process of cell wall or capsule synthesis, contain Golgi which appear to be in a high state of activity. This activity was ably demonstrated in the electron micrographs of Peyrière (20). Here, tetrasporogenesis in *Griffithsia flosculosa* involves the massive secretion of polysaccharides. A familiar pattern emerges with respect to the appearance of active Golgi in the red algae. According to Peyrière (21), the Golgi is polar, having a formative face (la face de formation) and a secretory face (la face de secretion). Endoplasmic reticulum and mitochondria always juxtapose the formative face, as do small membrane-bound vesicles. The secretory face produces dilated saccules which become increasingly rich in fibrillar material shown by histochemical techniques to be polysaccharide or mucopolysaccharide. The saccules give way to large vesicles which appear to move across the cell to the cell membrane, fuse with the membrane, and thus deposit the contained fibrillar material on the cell surface. The pattern here is a familiar one, and represents a verbatim visual representation of Northcote's physical flow scheme for polysaccharide secretion in plant cells (17).

This same pattern has been observed in many red algae, all presumably in the process of secretion, for instance in the apical cells of differentiating

Pseudogloiophloea (8), and in spermatia forming cells of *Bonnemaisonia* (22).

In capsule forming cells of *Porphyridium,* that is, cells actively involved in the synthesis, transport, and deposition of extracellular polysaccharides, this familiar pattern of active Golgi is also seen (8, 9). The evidence strongly suggests that the process of secretion of extracellular polysaccharides in *Porphyriddium* is a Golgi-mediated process.

However, evidence garnered from the integration of static election micrographs should not be over-interpreted. Experimental data is required to identify the compartments with the cells responsible for the synthesis, movement and deposition of red algal cell surface polysaccharides.

Biosynthesis of Red Algal
Extracellular Polysaccharides

Little is known about the biosynthesis of the extracellular DL-galactans of the red algae, despite the intense investigation of their structure and properties (1). Su and Hassid (23) showed that the water soluble extracellular polysaccharide of *Porphyra perforata,* upon hydrolysis, gave D-galactose, 6-0-methyl-D-galactose, 3,6-anhydro-L-galactose and sulfate ester in the molar ratios of approx. 1:1:2:1. Further, they isolated from the cells of *Porphyra* the sugar nucleotides UDP-D-galactose, GDP-L-galactose, UDP-D-glucose, GDP-D-mannose, and UDP-D-glucuronic acid, as well as 10 different nucleotides (24). They assumed the sugar nucleotides to be the precursors of the extracellular DL-galactan, and suggested the following sequence for its biosynthesis:

D-mannopyranosyl phosphate
$$\downarrow$$
GDP-D-mannose
$$\downarrow$$
GDP-L-galactose
$$\downarrow$$
DL-galactan.

The origin of the sulfate group in the *Porphyra* DL-galactan is also not known. However, Su and Hassid (23) suggest that the newly found (in *Porphyra*) nucleotide adenosine-3',5'-pyrophosphate may possibly activate the inorganic sulfate to form an "active sulfate donor" for an enzymic sulfation reaction. Freshly harvested *Chondrus crispus*, when suspended in a defined salts medium containing $^{35}SO_4^=$, removes the sulfate from the external medium and incorporates it into both the k- and λ-fractions of carrageenan (25). Similar experiments have been completed for *Porphyridium* (in this paper and 9). However, as with *Porphyra*, the pathway for the sulfation of the polysaccharides of both *Chondrus* and *Porphyridium* is unknown.

Acknowledgements

This work was supported by a grant (GB-18144) from the National Science Foundation. The author wishes to thank Mrs. Dawn Dodson, Mr. Samuel T. Groves and Dr. Thomas M. Jones for their assistance in this project.

References

1. D.A. Rees, *Advan. Carbohyd. Chem. Biochem. 24*, 267 (1970).
2. E. Percival, *The Carbohydrates*, IIb, W. Pigman and D. Horton, Eds. p. 537. Academic Press, N.Y. and London (1970).
3. R. W. Eppley, *J. Gen. Physiol.*, *41*, 901 (1958).

357

4. H. Hoogenhaut, *Naturwissenschaften*, 50, 456 (1963).
5. T. E. Brown and F. L. Richardson, *J. Phycol.*, 4, 38 (1968).
6. M. R. Sommerfeld and H. W. Nichols, *J. Phycol.*, 6, 67 (1970).
7. E. Gantt, M. R. Edwards and S. F. Conti, *J. Phycol.*, 4, 65 (1968).
8. J. Ramus, *J. Phycol.*, 8, 97 (1972).
9. J. Ramus and S. T. Groves, *J. Cell Biol.*, 54, 399 (1972).
10. T. M. Jones, Department of Plant Pathology, Cornell University; personal communication.
11. S. T. Groves, Department of Biology, Yale University; personal communication.
12. E. W. Yemm and A. J. Willis, *Biochem. J.*, 57, 508 (1954).
13. E. M. Livingston, et al., *Microchem. J.*, 1, 261 (1957).
14. C. Nalewajko and D. R. S. Lean, *J. Phycol.*, 8, 37 (1972).
15. J. D. Jamieson and G. E. Palade, *J. Cell Biol.*, 50, 135 (1971).
16 D. H. Northcote, *Endeavour*, 30, 26 (1971).
17. D. H. Northcote, *Ann. Rev. Plant Physiol.* 23, 113 (1972).
18. P. M. Ray, T. L. Shininger and M. M. Ray, *Proc. Natl. Acad. Sci. U.S.* 64, 605 (1969).
19. D. T. A. Lamport, *Ann Rev. Plant Physiol.* 21, 235 (1970).
20. M. Peyrière, *C.R. Acad. Sci. Paris, Ser. D 270*, 2071 (1970).
21. M. Pyerière-Eyboulet, *These, L'University de Paris XI, Centre D'Orsay* (1972).
22. J. Simon-Bicahrd Breaud, *C. R. Acad. Sci. Paris Ser. D. 274*, 1485 (1972).
23. J. C. Su and W. Z. Hassid, *Biochemistry 1*, 474 (1962).
24. J. C. Su and W. Z. Hassidd, *Biochemistry, 1*, 468 (1962).

25. F. Loewus, G. Wagner, J. A. Schiff, and
     J. Weistrop, *Plant Physiol.* 48, 373 (1971).

BIOSYNTHESIS OF CELLULOSE BY A PARTICULATE ENZYME SYSTEM FROM *ACETOBACTER XYLINUM*[*]

J. Kjosbakken[**]
and
J. Ross Colvin

Division of Biological Sciences
National Research Council of Canada
Ottawa, Canada

Abstract

A cell-free particulate enzyme system from *Acetobacter xylinum* which can form cellulose from uridine diphosphate glucose (UDPG) is also capable of forming compounds which are a combination between a lipid(s) and glucose, cellobiose and perhaps higher polymers of glucose. These compounds which are acid and alkali labile seem to be made prior to the formation of cellulose and their synthesis is repressed by both uridine monophosphate (UMP) and uridine diphosphate (UDP). These observations are interpreted by postulating that cellulose biosynthesis proceeds by way of a transient lipid-pyrophosphate-cellobiose compound.

---

[*] Issued as N.R.C.C. No. 12936

[**] National Research Council of Canada Post-doctoral Fellow, 1971-1973.

Introduction

One of the most interesting hypotheses in
cellulose biosynthesis over the last decade has been
the possible function of a lipid as a carrier for
glucose from the cell membrane to the site of
cellulose formation which is remote from the membrane. Colvin showed that, by incubating an ethanol
extract of an active culture with the appropriate
enzyme, fibrils were produced which were morphologically indistinguishable from cellulose microfibrils
and which, upon hydrolysis, gave rise to glucose (1).
This precursor was pruified extensively, but not to
homogeneity (2). Lately, Colvin and Leppard devised
an improved method for the purification of the
precursor (3). Colvin also showed that a similar
compound could be isolated from peas and oats but
not from beans (4).

More recently, similar models have been postulated for the biosynthesis of other extracellular,
carbohydrate-containing polymers. It was shown that
membrane-bound lipids function as sugar carriers in
the biosynthesis of peptidoglycan and of O-antigen
of the lipopolysaccharide of gram positive and gram
negative bacteria respectively (5, 6). The lipid
in both cases was shown to be a C_{55}-polyisoprenoid
alcohol. The sugars making up the repeating unit
of the polysaccharide are linked to this lipid by
a pyrophosphate bridge. Similar lipid intermediates
have been characterized in a few other bacterial
systems, and circumstantial evidence for this type
of intermediate has been found in other bacterial
as well as yeast, plant and mammalian cells (7-23).

Glaser described a cell-free particulate
enzyme system from *Acetobacter xylinum* capable of
cellulose formation when incubated with UDP-glucose
(24). If a lipid precursor of cellulose exists,
this system should also incorporate a small amount
of glucose from UDP-glucose into a lipid-fraction

and we have been able to confirm that this is the
case, thereby strengthening the hypothesis of a
lipid precursor of cellulose. Some properties of
the enzyme system and, in particular, some properties
of the lipids formed are reported here.

Materials and Methods

The preparation of the particulate enzyme from
Acetobacter xylinum by ultrasonic treatment of a
bacterial cell suspension, enzymatic synthesis of
cellulose from UDP-glucose and the incubation
mixtures for assay were essentially as described by
Glaser (24). The reactions were stopped and lipids
extracted by adding 20 vol. of a mixture of chloroform-methanol (2:1). The lipid extract was washed
four times to remove UDP-glucose. Cellulose was
prepared from the lipid-free residue by digestion
for five minutes in 4% sodium hydroxide at $100^{\circ}C$.
The insoluble residue was centrifuged, washed,
dried and counted as described by Glaser (24), except that the counting was done by a Beckman LS-250
liquid scintillation system. Estimation of the
radioactivity in aliquots of the lipid fractions was
by the same system.

Uridine diphosphate glucose-^{14}C, uniformly
labelled in the glucose moiety, was obtained from
New England Nuclear, Boston, Mass; (specific
activity, 227 mc/mmole).

Plates for thin layer chromatography (20 x
20 cm; precoated, Q1) were obtained from Quantum
Industries, Fairfield, New Jersey. Samples on the
plates were usually developed two-dimensionally,
first with a basic solvent, chloroform-methanol-
concentrated ammonia-water (65:25:2:2) and then with
an acidic solvent, chloroform-methanol-acetic
acid-water (170:75:25:6).

363

Paper chromatography was on washed Whatman No. 1 filter paper, usually with butanol-pyridine-water (10:4:4) as solvent. Visual detection of compounds was by alkaline silver nitrate (25).

Results

We confirmed that the particulate enzyme system produces labelled cellulose when incubated in the presence of UDP-glucose uniformly labelled with ^{14}C-glucose (24). Simulataneously, however, a small amount of radioactivity was incorporated into the lipid fraction. A comparison of the relative rate of formation of labelled lipid(s) and labelled cellulose is shown in Fig. 1. During the first five minutes a rapid formation of labelled lipids is observed, reaching a steady state level after about 10 minutes. In contrast, cellulose formation is almost proportional with time. However, initially more radioactivity is incorporated into lipid(s) than into cellulose, indicating that formation of the labelled lipid(s) preceeds formation of cellulose, as expected for an intermediate in the pathway between UDP-glucose and cellulose.

The labelled lipids were purified by a DEAE-cellulose column using an ammonium acetate gradient (26). The results are shown in Fig. 2. All radio-activity was eluted in one peak, showing either that only one labelled lipid is formed or, possibly, several with equal or very similar net charge. In addition, the labelled lipids are eluted as acidic phospholipids, in the same region as reported for a polyprenoid-pyrophosphate-sugar.

The original lipid extract and the labelled fraction from DEAE-cellulose were fractionated by thin layer chromatography in several solvent systems. The results showed that at least five labelled lipids

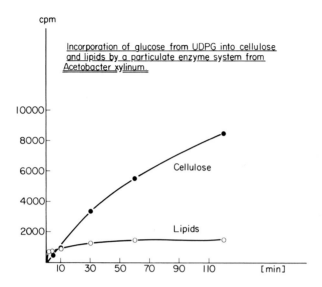

Fig. 1 Incorporation of uniformly labelled ^{14}C glucose from UDP-glucose into cellulose and lipids by a particulate enzyme system from *Acetobacter xylinum* as a function of time. Note that, although the final total incorporation into lipids is much less than for cellulose, initially the incorporation into lipids is substantially greater.

were formed, but they all migrated very close together, which means they must be chemically very similar.

The elution of labelled lipids from a DEAE-cellulose column by ammonium acetate suggests a pyrophosphate bond in the labelled lipids. To test this further the stability of labelled lipids towards weak acid and alkali was studied (27). Fig. 3 shows the effect of weak acid. In the untreated sample more than 90% of the total radioactivity was found to be lipid bound, while 95% of the radioactivity was found to be water soluble after treatment with 0.01 N H_2SO_4 at $100^{\circ}C$ for 10 minutes. Fig. 4 shows the effect of weak alkali. Again, 90%

365

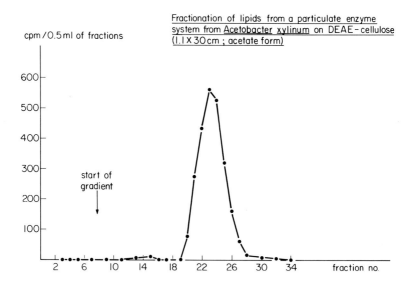

Fig. 2 Fractionation of the radioactive lipids extracted from a particulate enzyme system of *Acertobacter xylinum* which was incubated with UPD-^{14}C glucose, on a DEAE-cellulose column (1.1 x 30 cm; acetate form) eluted by a gradient of ammonium acetate (0-0.4 M in 99% methanol).

Acid hydrolysis of lipids isolated from a particulate enzyme system
from <u>Acetobacter xylinum</u>

	cpm	% in each phase
Untreated:		
Chloroform phase	6420	92
Water phase	600	8
	7020	100
Hydrolysed:		
Chloroform phase	330	5
Water phase	6060	95
	6390	100

Fig. 3 Distribution of radioactivity of labelled lipids between chloroform and water phases before and after acid hydrolysis of the lipid extract.

Alkali treatment of lipids isolated from a particulate enzyme system
from <u>Acetobacter xylinum</u>

	cpm	% in each phase
Untreated:		
Chloroform phase	6210	90
Water phase	700	10
	6910	100
Hydrolysed:		
Chloroform phase	650	10
Water phase	6020	90
	6670	100

Fig. 4 Distribution of radioactivity of labelled lipids between chloroform and water phases before and after alkaline hydrolysis of the lipid extract.

of the radioactivity was found to be water soluble after treatment with 0.1 N LiOH at room temperature for 30 minutes.

The labelled substances which were released by acid hydrolysis were characterized by paper chromatography and by gel filtration. Three radioactive compounds could be detected by paper chromatography: one migrating as glucose, one as cellobiose and one staying at the origin. The last compound(s) is not glucose-1-phosphate but it could be higher oligosaccharides of cellulose. The manner of elution of the water-soluble compounds from a Sephadex G-10 column is shown in Fig. 5. Confirming the results with paper chromatography, two compounds were eluted from the column with the same elution volume as authentic standards of glucose and cellobiose. A small amount of radioactivity was eluted in front of cellobiose, indicating higher oligosaccharides of cellobiose.

These results demonstrate that a particulate enzyme system from *Acetobacter xylinum* can produce labelled cellulose from radioactive UPD-glucose and, at the same time, a small amount of labelled lipids. In these lipids, glucose, cellobiose and perhaps

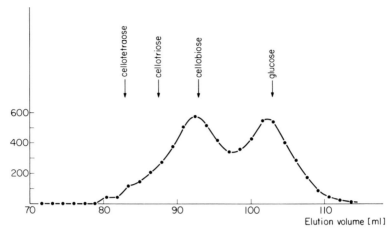

Fig. 5 Fractionation of water soluble fragments from the acid hydrolysis of labelled lipids on Sephadex G-10 eluted with 0.025 M ammonium acetate. The arrows indicate the elution volume of authentic standards of glucose, cellobiose, cellotriose and cellotetrose.

higher molecular weight oligosaccharides are bound weakly, possibly by a pyrophosphate bridge. These lipid compounds may be similar to the poly-prenoid-pyrophosphate-sugars shown to be intermediates in the biosynthesis of several other extracellular polysaccharides. If a lipid-pyrophosphate-cellobiose is the precursor of cellulose the following reactions may be postulated; first, glucose-1-phosphate is transferred from UDP-glucose to a lipid-phosphate, forming a lipid-pyrophosphate-glucose. In a second reaction, glucose is transferred from UDP-glucose, forming a lipid-pyrophosphate-cellobiose. In the next step, cellobiose is transferred from the lipid to the growing tip of the cellulose microfibril. The lipid after dephosphorylation is then back at the start of the cycle. According to this hypothesis UMP from the first and UDP from the second reaction would be end products, and we have shown that an excess of each of the two nucleotides strongly inhibits the formation of labelled lipids.

Discussion

The foregoing results show that the cell-free particulate enzyme system from *Acetobacter xylinum*, which was first described by Glaser (24), is capable of forming not only cellulose but also several compounds which are a combination between lipid(s) and glucose, cellobiose and perhaps higher polymers of glucose. These lipid-sugar compounds which are labile both to acid or alkali seem to be formed prior to cellulose and their formation is inhibited by adding either UMP or UDP to the system. The observations are consistent with the suggestion that cellulose biosynthesis proceeds by the prior formation of a lipid-pyrophosphate-cellobiose. The cellobiose residue is then transferred to the end of a pre-existing polyglucosan chain at the tip of a microfibril. Up to now, the model whereby glucose molecules are incorporated one by one into the growing tip of the cellulose microfibril seems to have been favoured. However, as Dr. Blackwell of the Case-Western Reserve University suggested to us, this is not a simple reaction stereochemically since every second glucose residue of the polyglucosan chain occupies a plane which is rotated by 180° compared to its nearest neighbor. By postulating cellobiose as the basic unit which is incorporated,one is not required to face this problem, and in fact cellobiose, not glucose, may be regarded as the repeating unit of cellulose in terms of stereochemistry.

At the moment, we have evidence from electron microscopy experiments that the lipid extract from the particulate enzyme system in fact contains a compound which on mixing with an enzyme produces cellulose microfibrils. So far, however, we cannot conclude that this precursor of the microfibrils and our postulated lipid-pyrophosphate-cellobiose is the same compound. To prove this we have to isolate the labelled lipid(s); work toward this end is now in progress.

References

1. J. R. Colvin, *Nature* *183* 1135 (1959).
2. A. W. Khan and J. R. Colvin, *J. Polymer Sci.* *51* 1 (1961).
3. J. R. Colvin and G. G. Leppard, *J. Polymer Sci.*, *Part C*, *36* 417 (1971).
4. J. R. Colvin, *Can. J. Biochem. and Physiol.* *39* 1921 (1961).
5. Y. Higashi, J. L. Stronimger and C. C. Sweeley, *Proc. Nat. Acad. Sci. U.S.* 57 1878 (1967).
6. A. Wright, M. Dankert, P. Fennessey and P. W. Robbins, *Proc. Nat. Acad. Sci. U.S.* 57 1798 (1967).
7. W. Tanner, *Biochem. Biophys. Res. Comm.* 35 144 (1969).
8. R. Sentandreu and J. O. Lampen, *FEBS Letters* *14* 109 (1971).
9. S.S. Alam and F.W. Hemming, *FEBS Letters* *19* 60 (1971).
10. R. M. Barr and F.W. Hemming, *Biochem. J.* *126* 1203 (1972).
11. H. Kauss, *FEBS Letters* *5* 81 (1969).
12. L. J. Douglas and J. Baddiley, *FEBS Letters* *1* 114 (1968).
13. D. Brooks and J. Baddiley, *Biochem. J.* *113* 635 (1969).
14. R.G. Anderson, H. Hussey and J. Baddiley, *Biochem. J.* *127* 11 (1972).
15. I. C. Hancock and J. Baddiley, *Biochem. J.* *127* 27 (1972).
16. H. Hussey and J. Baddiley, *Biochem. J.* *127* 39 (1972).
17. N. H. Behrens and L. F. Leloir, *Proc. Nat. Acad. Sci. U.S.* *66* 153 (1970).
18. N. H. Behrens, A. J. Parodi, L. F. Leloir and C. R. Krisman, *Arch. Biochem. Biophys.* *143* 375 (1971).
19. N. H. Behrens, A. J. Parodi and L. F. Leloir, *Proc. Nat. Acad. Sci. U.S.* *68* 2857 (1971).
20. C. L. Villemez and A. F. Clark, *Biochem. Biophys. Res. Comm.* *36* 57 (1969).

21. M. Scher, W. J. Lennarz and C. C. Sweeley, *Proc. Nat. Acad. Sci. U. S. 59* 1313 (1968).
22. K. Takayama and D. S. Goldman, *J. Biol. Chem. 245* 6251 (1970).
23. F. A. Troy, F. E. Frerman and E. C. Heath, *J. Bio.. Chem. 246* 118 (1971).
24. L. Glaser, *J. Biol. Chem. 232* 627 (1958).
25. S. M. Martin, *Chem. and Ind.* 823 (1957).
26. M. Dankert, A. Wright, W. S. Kelley and P. W. Robbins, *Arch. Biochem. Biophys. 116* 425 (1966).
27. K. Nikaido and H. Nikaido, *J. Biol. Chem. 246* 3912 (1971).

Note Added in Proof (November 16, 1972)

After this paper was presented and the article accepted for publication, we learned of the work of Dankert, Garcia and Recondo which was published as a communication to a symposium in San Carlos de Bariloche, Argentina, November 1971 (*Biochemistry of the Glycosidic Linkage; An Integrated View.* Eds., R. Paras and H. G. Pontis; Academic Press, New York, p. 199, 1972). Working with a butanol extract of EDTA-treated cells rather than with a cell-free enzyme preparation, these workers reached essentially the same conclusions as we have done independently. To us, the correspondence of results and conclusions is a source of intense satisfaction.

SUBJECT INDEX

A

A-fraction of IAA esters, 301
  characterization, 306-312
  purification, 304-306
Acer pseudoplatanus, see sycamore maple
Acer saccharum arabinogalactan, 135
Acetabularia mannan, 251-252
Acetoacetic acid decarboxylation, 74
Acetobacter xylinum cellulose, 209, 293,
  361-371
ADP-glucose, 31
Alcian Blue, 212, 337, 338
Alditol acetate(s)
  methylation, 119
  preparation, 120
  separation, 119
Aldolase, 11-13
Algae, 207-255, 333-357
  haptophycean, 209
Amplexus, 215
Ampullariella digitata carboxy-lyase, 79
Anthrone assay, 343, 353
  glutaraldehyde interference, 208
Apiobiose, 88
  chromatography, 92
Apiofuranose, acid lability, 106
Apiogalacturonan(s), 85-94
  biosynthesis, 90-93
  degradation, 88
  extraction, 87
  structure, 88
Apiose, 85-94
  chromatography, 92
Apiose (cyclic 1:2)-P, 89
Arabinan
  branching, 108, 133
  in *Acer* pectin, 130, 133, 134
  in lemon peel, 107, 135
  in soybean, 107, 135
  link to rhammogalacturonan, 134
  structure, 102

Arabinofuranose, 97
  acid lability, 106
  from arabinopyranose, 80
  origin, 101
Arabinogalactan
  in *A. saccharum,* 135
  3,6-link in *Acer* pectin, 130, 135
  link to serine, 136
  structure in *Acer,* 136, 137
  type I, 105, 107
  type II, 97, 99
Arabinopyranose, 97
  conversion to furanose, 80
  4-epimerization, 101
Arabinose
  biosynthesis, 69-82
  conversion to UDP-arabinose, 78
  in *Acer* xyloglucan, 124
  in pollen tube pectin, 18
Arabinosyl hydroxyproline in extension,
  149-161
Auxin, 141, 142
  bound, 297-314
Avena sativa, cellulose biosynthesis, 268

B

B-fraction of IAA esters, 301-303
Barley
  pectin formation, 7
  seedling weight, 56-57
Betulaprenol phosphate, 293
Biogenesis, *see* biosynthesis
Biosynthesis of,
  apiogalacturonan, 90-93
  arabinose, 69-82
  cellulose, 207-255, 259-293, 361-371
  fibrillar polygalacturonic acid, 326-327
  glucuronic acid, 51
  hydroxyproline-rich glycoprotein,
    165-173

D

Daucus carota, 317
  cell wall fibers, 323
  hydroxyproline-rich glycoprotein, 166
Decarboxylation, 74
Diborane reduction of methylated
  galacturonosides, 131
Dihydroxyacetone P, 11-12
Dipyridyl inhibition, 168, 170
dTDP-galacturonic acid, 79
dTDP-glucose oxidoreductase, 11

E

Electron microscope studies of
  capsular polysaccharide, 341-342, 352
  cellulose biosynthesis, 280, 369
  cellulosic glycoprotein, 211-246
  polygalacturonic acid fibers, 317-319
  stylar transmitting tissue, 196-198
Endoglucanase, 126, 141
Endopolygalacturonase of *C. lindemuthianum,* 124
  in galacturonan degradation, 131, 140
  on *Acer* pectin, 130
4-Epimerases, evolution, 80-81
Erythritol from Smith degradation, 271, 275
N-ethyl pyridium chloride, 252
Extracellular polysaccharides, 119, 177-188,
  253, 328-329, 333-379
Extensin, 149-163

F

Fucose, 99, 124
Fucosylgalactose sidechains, 126
Fucoxyloglucan, 99, 100

G

Galactan, 102, 108
  in *Acer* pectin, 130, 133, 134
  in soybean, 134
  sulfated, 334
Galacto-glycopeptide amino acid sequence,
  152

Galactofuranose, 80
Galactopyranose, 97
Galactose in
  Acer xyloglucan, 124
  capsular polysaccharide, 337
  glycopeptides, 169
  scale fraction A, 246
  serine-linked structures, 154, 169
Galactosyl hydroxyproline, 153
Galactosyl serine, 159
  in extensin, 149-163
  in galacto-glycopeptide, 152
Galacturonan, 100, 102, 104
Galacturonic acid in
  intercellular substance, 202
  Ipomoea cell fibers, 321
  plant cell wall fibers, 316
  pollen tube wall pectin, 18
Galacturonorhamnan, 100
Galacturonosyl-(1-2)-rhamnose, 132
GDP-galactose, 356
GDP-glucose, 263, 264, 268, 271, 275-280,
  356
GDP-mannose, 275-280, 356
Glucan, 251, 263, 264, 267, 268
  IAA ester of, 297-314
  pollen tube, 251
  synthetase(s), 209, 253, 280
Glucomannan, 97, 98, 275
Glucose
  conversion to glucuronic acid, 3-4, 51
  conversion to xylose, 72
  cyclization to *myo*-inositol, 3-6
  in A-fraction of corn, 310
  in capsular polysaccharide, 337
  in lily pistils, 177
Glucose-1-^{14}C conversion to
  exudate, 184-188
  pectin, 4-6, 20-25, 196
Glucose-U-^{14}C metabolism in lily pistils,
  188
Glucose-6-P cycloaldolase, 3-4, 8-13
  bound NAD$^+$ in, 10-11
  occurrence, 8
  purification, 8
Glucosyl transferase localization, 280, 282
Glucuronic acid, 97
  from *myo*-inositol, 3, 31, 51
  in intercellular substance, 202